READING PHYSICS

READING PHYSICS

A GUIDE TO UNDERSTANDING CLASSICAL MECHANICS WITHOUT MATHEMATICAL EXPRESSIONS

JAE JUN KIM, PH.D.

Universal-Publishers
Irvine • Boca Raton

Reading Physics: A Guide to Understanding Classical Mechanics
without Mathematical Expressions

Universal Publishers, Inc.
Irvine • Boca Raton
USA • 2023
www.Universal-Publishers.com

ISBN: 978-1-62734-428-9 (pbk.)
ISBN: 978-1-62734-429-6 (ebk.)

Typeset by Medlar Publishing Solutions Pvt Ltd, India
Cover design by Ivan Popov

Library of Congress Cataloging-in-Publication Data

Names: Kim, Jae Jun, 1981- author.
Title: Reading physics : a guide to understanding basic classical mechanics
    without mathematical expressions / Jae Jun Kim, PH.D.
Description: Irvine, California : Universal Publishers, Inc., 2022.
Identifiers: LCCN 2022043305 (print) | LCCN 2022043306 (ebook) |
    ISBN 9781627344289 (paperback) | ISBN 9781627344296 (ebook)
Subjects: LCSH: Physics. | Mechanics.
Classification: LCC QC23.2 .K56 2022 (print) | LCC QC23.2 (ebook) |
    DDC 530--dc23/eng20221121
LC record available at https://lccn.loc.gov/2022043305
LC ebook record available at https://lccn.loc.gov/2022043306

To Hye Sun, Arin and Joowon

Abstract

This book was written to help college students understand physics without complicated math. Each year, thousands of college students pursuing business and humanities degrees find themselves taking a course in introductory physics. But many will have serious trouble solving physics problems because they don't have enough experience using mathematical equations. Understanding physics is challenging without a strong math background but not impossible. Unlike other introductions to physics, this book explains the basic concepts in classical mechanics with a minimal number of mathematical expressions, so new students can spend their time learning physics, not math. The result is not only a better understanding of physics but possibly a better grade. This study guide covers the three aspects of classical mechanics: basics of motion, rules for gravitational interaction among two or more objects, and rotational motion. The bottom line is that this book was written to help students better understand the mathematical parts of undergraduate classical mechanics so that they can concentrate on learning physics, not math.

Table of Contents

Prologue

"It was almost 50 percent."

Yes, it was almost always about 50 percent. What was that about? I had been teaching general physics courses, and it was the average percentage of students having trouble understanding basic mathematics. Even considering that most students in the course pursued their academic interests in business and humanities, seeing that percentage was an alarm to me. I thought about it hard. What was going on? Was there something I could do to at least mitigate the issue? In the end, having such thoughts led me to do something to help them understand the classical mechanics better. So, I began writing my "we read physics" series.

I have no doubt that mathematics is a useful and highly effective tool to play with when studying physics. To some extent, it is necessary to grasp basic mathematical concepts for those who consider majoring in physics or other engineering sciences. But you know what? In the end, those who major in physics need to utilize it over and over, so mathematics is going to be a part of their life. They will get it eventually.

But what about students not majoring in physics or other engineering science? What about students preparing for medical school or other professional school exams without having that sort of scientific background? Are they going to have sufficient time to grasp the mathematics part of the classical mechanics? How about motivating the students not majoring in physics and encouraging them to go over the materials being covered in undergraduate classical mechanics if they don't have enough time to fully understand them?

My experiences taught me a lesson: the reality is brutal. It really is. Most students do not even have opportunities to read and focus on the physics parts in general physics. They just do not have that time. So, what do they do instead? Yes, they spend their time on understanding the mathematical side of the physics instead. What is the consequence? Well, they end up working on their assignments using the mathematical equations without understanding why they do so much and do it that way. Then, when they are asked to work on some projectile motion, they get lost and things get worse and worse by the time they meet Newton's laws.

Coming back to the percentage, almost 50 percent of students in my classes had that issue. Yes, it was about 50 percent of them. The bottom line is that I could not simply ignore the issue but needed to take some action instead and deal with it and try to come up with a resolution, a practical resolution that we can all consider for students studying physics in the future.

Let us think about time for a moment. When is the best time to initiate the rescue? After having their first test in class, it might be too late for instructors to initiate their motivation back again. For many students, they simply follow what is being covered in class and get lost further. Then, even before they realize it, the semester is over. That is the end of the story.

This causes a few issues. First, students might think physics is just about using equations and solving problems, even without understanding the equations. I am not saying that is not important, but if that is all they study and get lost, we need to think about our traditional approaches. Furthermore, all they are going to remember is the mathematical part of physics, not the physics part of it. Combining these two issues, students end up having no fun taking their first physics course. In particular, the latter can cause more serious issues for those studying and preparing for an admission test. The test is mainly about reading a paragraph on physics topics and working on their problems. It is not only about playing with numbers but also about understanding the materials instead. So, those who spent too much time focusing on the mathematical parts of classical mechanics are going to have hard time grasping the core concepts of the questions in the exams. What can we do to resolve these issues?

Reading Physics is written to help them out, particularly those who struggle with understanding the fundamentals of vectors and trigonometric functions and for those eagerly looking for a way to understand the physics part of physics in classical mechanics better. Some may say it is not practical or possible to do so. However, I truly believe that reading this book before or while taking a course in physics is going to help students get motivated to study further and get them prepared to understand the physics side of physics. In terms of practicality, that is better than just getting lost in the middle and not studying physics anymore. It is better than not studying at all. We, as a physics community, need to be more realistic in our approach toward educating students.

One more thing: the author tried to minimize introducing numerical equations in this book. Almost everything is written in plain sentences. Why is that? It gives you an opportunity to read it through and then read it again without getting interrupted by the mathematics when you are having trouble understanding them. Think of how many times students ended up getting lost in understanding a single equation and then ended up closing their book and never looking back. It happens more than you think. I did my best to avoid that tragedy.

Does that mean that reading this book is going to be sufficient for students to gain an overall knowledge of classical mechanics and that this is going to be an ultimate resource for them preparing for their medical school admission test? Probably not. I would say that this is written as a starting point for your studying physics. After reading this book, my recommendation is to study further by reading books with many mathematical expressions. I strongly encourage you to study the mathematical portion of the classical mechanics using other books or notes. You can go over some video lectures on your own, for instance. When you do so, you will have a much less hard time if you read this book first, understand the fundamentals, and then study or take lectures in classical mechanics.

In summary, I truly hope that this book helps students who are having trouble to understand the mathematical parts in undergraduate classical mechanics. I do strongly hope that this book is going to function as a bridge through which students, who are having issues understanding classical mechanics, are going to better understand the physics side of it. It will be even better if students get inspired to study physics further and think like physicists. Thank you, and welcome to the world of classical mechanics.

CHAPTER 1

Kinematics

We study mechanical motion associated with a single object in a system using some physical quantities.

Day 1
Everything is relative

Jae: I drove my car 100 miles per hour
yesterday.
Adam: Well, that could be dangerous.
Chris: Wait a second, *with respect to what*
did you drive your car at that speed?

What a lovely day. It was lovely to go outside. After this chain of thoughts, I decided to go on a trip to New York. I live in Columbia, the capital of South Carolina, so I decided to drive on a highway to get to Charlotte, a city in North Carolina, and then take a plane to get to New York from there. To tell you the truth, there was a physics conference that I wanted badly to attend, so I needed to get there anyway.

But you know what? I was so excited that I drove too fast. I drove 100 miles per hour on the highway. And, to make the case more interesting, everyone thought that it was such a lovely day and that led to a police officer driving a car at the same speed on the opposite side. In other words, a police car was driving 100 miles per hour on the opposite side. Question: what was the speed of the police car? Was it 100 miles per hour? Or 200 miles per hour? Or some other speed?

Now we are getting to the real story. If you decided to go with the first choice as your answer, that is great, although we may just need one little piece of information to fully describe the motion in a more practical manner. In any case, you can start reading this book in a joyful manner.

On the other hand, if you went with the second choice as your answer, that is awesome. Why? Because you might have been thinking about the "relativeness" associated with physical quantities such as velocity. Hint: think about the title of this lesson. In any case, you are also ready to have some joyful time studying classical mechanics. It is waiting for you!

Or, if you decided to go with the last choice, the "some other speed," something that you come up with on your own, that is truly and really awesome. Why? Because making that choice tells me that you do understand the fundamentals of relativeness. Or, as a best-case scenario, you can come up with an answer on your own. Just in case, if your answer happened to be something like, "It depends on how we set a reference," then, well, you really got me there. I can tell you that you may jump on to the next lesson and read the rest of the book thoroughly with joy and cheers. Why? Because coming up with such an answer tells me that you already understand important aspects of the fundamentals of the relativeness associated with physical quantities. Believe it or not, all the lessons in this book heavily rely on your understanding of "relativeness." Everything is relative. You will see the importance of understanding the point over and over when you end up studying Newton's laws in dynamics. Figure 1.1 illustrates this point. In essence, you are a being in motion, but depending on how it is being measured, the degree of your being in motion is going to be recorded differently.

Coming back to the main subject, here is the bottom line: we should have a reference with which a physical quantity can be measured or calculated. In addition, the numerical value associated with a measured quantity "depends on" where we set the reference. There are no exceptions. It may sound simple, but it is important to grasp the main point here, so I encourage you to digest the essence of the relativeness before moving on to the next lesson. Again, we need to have a reference to quantify physical quantities. Depending on where the reference point is, the size of the measured physical quantity could be different. For instance, coming back to the case of driving a car on a highway, if there happened to be someone standing still on the road measuring the speed of the police car, then the person is going to think the speed of the police car is 100 miles per hour. However, if you drive your car 100 miles per hour and measure the speed, you are going to think the police car is approaching you going 200 miles per hour. If it is hard to imagine the point here, think about your experiences of driving a car on a highway. Do the cars on the other side of the highway seem to be moving faster than the cars that are on the same side of road? Figure 1.2 might help you understand the point better.

You drive your car 100 miles per hour with respect to someone standing still. At the same time, when a police car is driving the same speed as you, coming from the opposite direction, you might think that the police car is approaching you at 200 miles per hour. The speed depends on how and where we set a reference.

Well, it sounds very odd, doesn't it? You may think that a mechanical motion can be quantified with an absolute value. Well, not in reality. That is what makes studying physics and engineering science so interesting. As you read this book and go over some practice questions, you are going to understand the fundamentals of the relativeness more clearly.

The next lesson is going to be about vectors in physics. If you do not fully understand the relativeness clearly at this moment, that is okay. Let us move on for now, and you can come back to this lesson once you go over the kinematics part first. It will help you understand the essence of relativeness more clearly and provide you with an opportunity to think about references in physics.

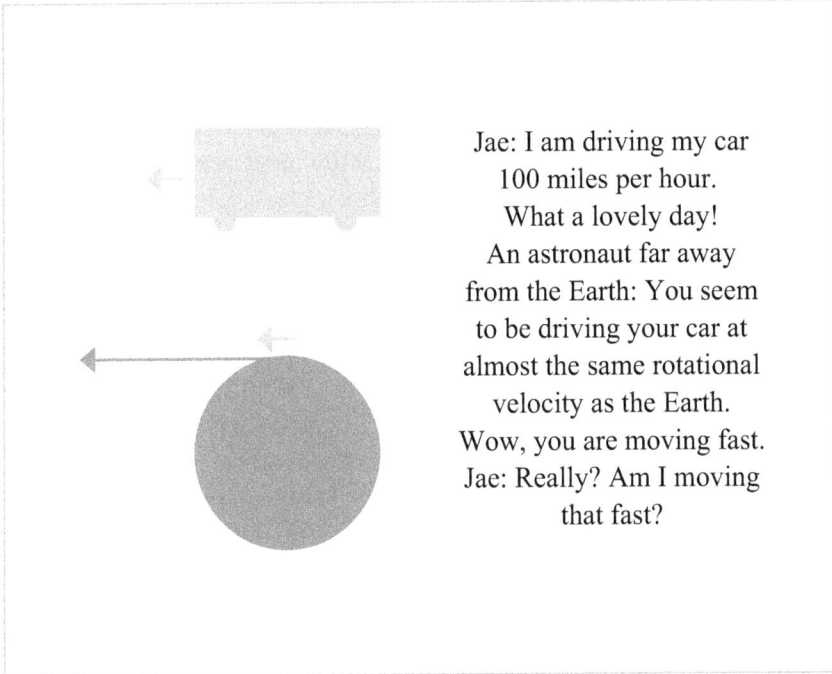

Jae: I am driving my car
100 miles per hour.
What a lovely day!
An astronaut far away
from the Earth: You seem
to be driving your car at
almost the same rotational
velocity as the Earth.
Wow, you are moving fast.
Jae: Really? Am I moving
that fast?

Figure 1.1: As the name of the lesson suggests, everything is relative in physics. You think that you are driving your car 100 miles per hour. Strictly speaking, when we say something like that, we do need to specify a reference. The 100 miles per hour as the velocity is probably measured with respect to the surface of the Earth. Question: what if the speed is measured by someone living far away from Earth? How is the person going to think about the velocity of you driving the car? Answer: The speed is going to be measured with a different number. Physical quantities such as speed and velocity depend on where and how we set a reference. Think about the size of the arrow in light gray on the top and the bottom part of the figure. Why are they drawn like that?

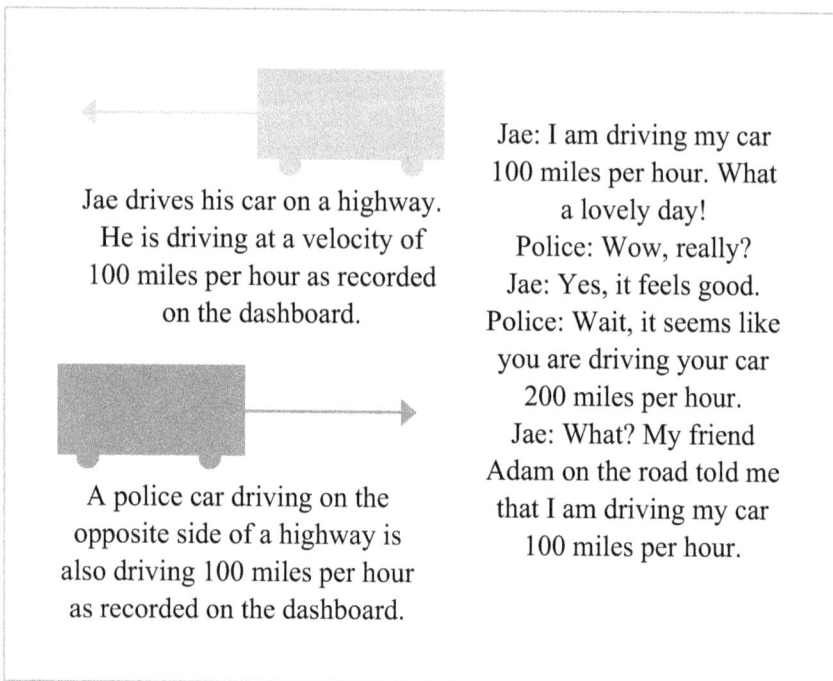

Jae drives his car on a highway. He is driving at a velocity of 100 miles per hour as recorded on the dashboard.

A police car driving on the opposite side of a highway is also driving 100 miles per hour as recorded on the dashboard.

Jae: I am driving my car 100 miles per hour. What a lovely day!
Police: Wow, really?
Jae: Yes, it feels good.
Police: Wait, it seems like you are driving your car 200 miles per hour.
Jae: What? My friend Adam on the road told me that I am driving my car 100 miles per hour.

Figure 1.2: This further illustrates the relativeness of physical quantities and emphasizes the importance of specifying a reference when measuring quantities to describe motions. With respect to someone standing still on the road, Jae drives his car 100 miles per hour. However, with respect to a police car driving on the opposite side of the lane but at the same speed, the police officer is going to think you are driving 200 miles per hour. The speed that we measure depends on where and how we set a reference. It always does.

Remember that grasping the essence of the lesson may not appear to be important when studying classical mechanics. It may appear to be so since we assume the surface of the Earth to be our reference when working on many practice questions in kinematics, but it may not be the case when working on dynamics, where a reference can be set differently and a final quantity may be different depending on the reference. If you do not clearly understand this lesson, you can go ahead and study the kinematics part, but I strongly encourage you to come back to this lesson one more time, read it through, and ensure that you understand this lesson before going on to study dynamics. You will have a much easier time when taking an introduction to classical mechanics later.

Problems:

Describe whether you can measure a physical quantity without a reference point. If you can, why is that? If you can't, why not?

You drive a car 100 miles per hour on a highway with respect to someone standing still on the side of the highway and a policeman drives his or her car on the opposite side of the lane with a hundred-mile-per-hour velocity. Calculate the size of the velocity of the police car with respect to "your car."

You drive a car at a hundred miles per hour on a highway with respect to someone standing still on the side of the highway and a policeman drives his or her car on the opposite side of the lane with a 100-mile-per-hour velocity. It is a rainy day, and the raindrops move down vertically at 100 miles per hour. Calculate the velocity of the raindrop with respect to your car. Is the velocity going to be same when measured by someone standing still on the side of a highway?

Day 2
Vector

"Vector is a quantity with a magnitude
and a direction, and scalar is a quantity
with a magnitude only. In general,
a mechanical motion can be represented
by a vector. It needs to be a vector,
not a scalar."

Back in December 2009, I had a trip to New York, one of the major cities in the United States. I went there to attend a meeting, and I enjoyed the nice weather. While staying there, I realized that New York is such an interesting city and different from where I had been living for many years. In any case, I learned a lot of physics while attending the meeting, but, interestingly, I learned a lot more about teaching physics when returning to my hometown. Here comes the story.

I had to take three flights back to Columbia, my hometown city in the state of South Carolina; the first flight was from New York to Charlotte, the second one from Charlotte to Atlanta, and the last one from Atlanta to Columbia. There was no issue with the first flight going from New York to Charlotte. I enjoyed the flight. But, as you may know, things happen. While waiting in an airport in Charlotte, I was told that they were having a small issue with the fuel tank in the plane. The tank was leaking. In the end, fixing the leak was taking too long, so the agents began to offer lodgings to the customers. I was one of them. After waiting for about 10 minutes or so, I spoke with one of the customer service agents. The agent looked through my flight itinerary and started to say something, "Why are you not going to ...?"

The agent did not finish the sentence, but it did not take much time for me to guess what they wanted to say. With all probability, the agent wanted to say the following: "Why are you not driving directly from here to your hometown?" Let us think about that. Why was the agent about to say something like that to me?

Just driving from the city of Charlotte to the city of Columbia, it would only take about 90 minutes, but taking two more flights, it would take a lot more than 1 hour and 30 minutes. The flight from Charlotte to Atlanta would take about 2 hours. There would be waiting time at the Atlanta airport, and then the flight from Atlanta to Columbia alone would take about 1 hour or so. Therefore, it would take at least 10 hours of my time to get to Columbia. The customer agent immediately noticed that fact when going over my itinerary. Figure 2.1 might help you understand the situation better; in terms of the starting and the ending position, driving down to Columbia or taking two flights would realize the same goal. At the same moment, I realized that this would be a good example to explain to the students in my class.

Let us go back to the issue again and think about some relevant information to my going from Charlotte to Columbia as going over Figure 2.1. What would be my travel distance if I chose to drive from Charlotte to Columbia? Say it would be about 100 miles. Now, what happens if I choose to take the two flights? The distance from Charlotte to Atlanta is going to be about 400 miles and from Atlanta to Columbia is 300 miles; so, the total distance would be about 700 miles. What a huge difference.

In the end, either way, the fact of the matter was that I would be moving from Charlotte to Columbia, but I needed to travel 100 miles in the first situation and 700 miles in the second one. Again, either way, my departing place and destination would be the same, but the total travel distance is different by about 600 miles, depending on which route I take. What is going on?

Here comes the important lesson: direction matters when going from Charlotte to Columbia or from one place to another in general, and depending on which direction I chose to go, the total distance could be different. There are an infinite number of possibilities that we can think of as far as choosing a route goes. That means that when describing motion associated with an object, we need to choose a single route, and both direction and magnitude need to be specified.

Question: how do we do that? We introduce a "vector," an important mathematical quantity that we use a lot when studying physics.

> When describing the motion of an object, both magnitude and direction matters, so we need to introduce a vector, a quantity that carries information regarding both a magnitude and a direction.

In mathematical terms, you can think of a vector as,

$$\text{Vector} = \{\text{Magnitude, Direction}\}$$

Going back to the story, in the former case, if I took the shortest route, going directly from Charlotte to Columbia, it would get me to my hometown in the shortest time and that could certainly be represented as a vector. The magnitude is going to be the distance between the two cities, and the direction is going to be the direction that goes directly down from Charlotte to Columbia.

Now, let us go over Figure 2.1 one more time. It is going to be simply drawing an arrow as a straight line, where the tail is at Charlotte and the head is at Columbia, and where the length of the vector represents the distance between the two cities. What is important to remember is that we always need both the size and the directional information to represent the shortest route between two cities. If the destination is not Columbia, but say, Chicago, then a different vector is needed because the direction of moving from Charlotte to Chicago is going to be different from my moving from Charlotte to Columbia. Furthermore, the moving distance is going to be different too.

You may ask the following question: what if I chose to head to Atlanta first and then from Atlanta to Columbia instead? Then the direction of my moving from Charlotte to Atlanta would not be aligned with that of moving directly from Charlotte to Columbia, so in the end I would need to take another flight. I needed to take two flights, each of which take me from one place to another, but, in the end, taking the two flights together gets me to my destination. So, here we learned another important lesson: the motions can be represented by vectors. On top of that, we can also add them or subtract one from the other.

My going from Charlotte to Atlanta can be represented as a vector, the distance as the length of the vector, and the direction representing the arrow of the vector. We can do the same for my going from Atlanta to Columbia. Now, let us think for a moment. Taking both "vectors" together, as mentioned before, I ended up arriving in my hometown, which is the same as going directly from Charlotte to Columbia.

Again, that is all that matters when describing a motion. So, the two vectors can be added and can be represented as the vector of directly going from Charlotte to Columbia. Summarizing it,

> We can add vectors: a vector representing
> going from Charlotte to Atlanta and
> another vector representing going from
> Atlanta to Columbia, and the added
> vectors can be represented by a single
> vector that represents my going directly
> from Charlotte to Columbia. We can do
> the same when subtracting vectors too.
> You just flip it.

Just out of curiosity, let us think about going from Columbia to Charlotte. Can that be represented as a vector? Yes, it certainly can because it can be represented as a mechanical motion. It can be represented by a vector of going from Charlotte to Columbia but with a small difference, and that is "the head of the arrow" being in Charlotte, not in Columbia. What does that mean? Answer: That means that we just flip the vector to the opposite direction so that the head of the vector as an arrow represents the destination, and the tail represents the departing place. That is represented as a vector that represents going from Charlotte to Columbia; mathematically, you simply add a negative symbol in front.

So far, I described a few properties of a vector using my going from Charlotte to Columbia by two different routes. Again, a vector has both a magnitude and a direction. Vectors can be added, and vectors with a minus symbol represent a quantity that has the same magnitude but an opposite direction. There are more intriguing and interesting properties associated with vectors, but let us move on. You may find additional properties in other literature if you are interested. I think I have said enough about the importance of understanding vectors when studying classical mechanics. Let us move on to the next subject.

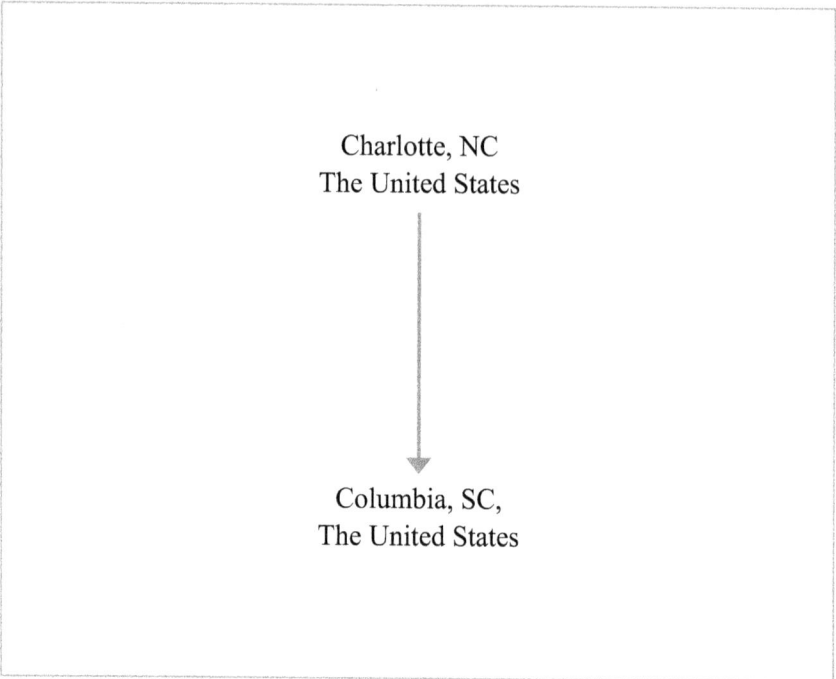

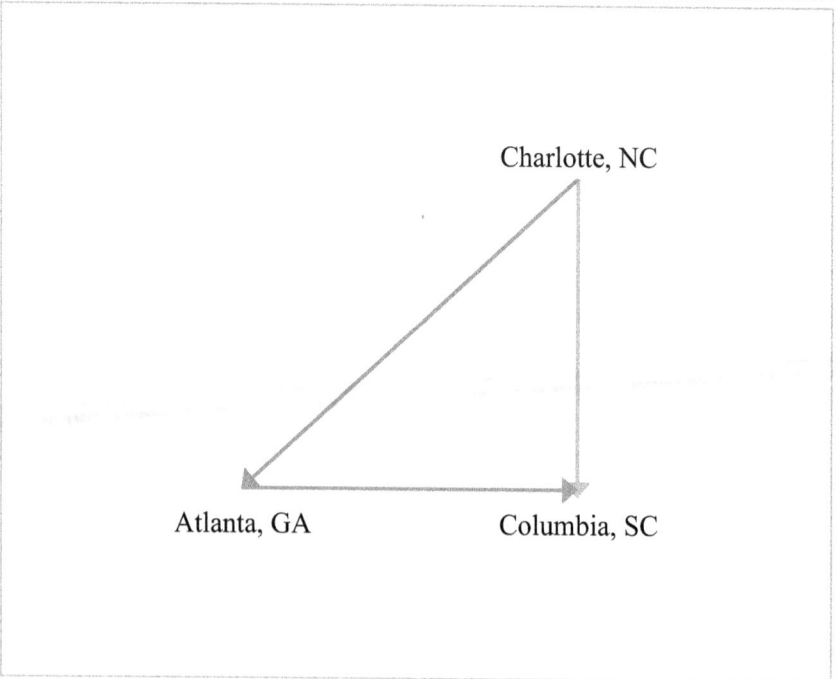

Figure 2.1: Believe it or not, the figure on the top and bottom is all
that we need to understand the properties associated with vectors here.
Among the many different routes I can take to get to the destination,
two different routes are shown to go from Charlotte to Columbia: one
where I could directly go from Charlotte to Columbia, and another where
I could go to Atlanta first and then go to Columbia next. The former is
shown on the top, and the latter on the bottom. For both, the start and the
end position happen to be same. The departing place was Charlotte, and
the destination was Columbia. The "effective" distance, or what is called
"displacement" in physics, which you will study later, of the routes is
the same no matter which route is taken. It is just that the "distance" in
the former is not the same as in the latter. You go to Atlanta first and then
arrive at Columbia next. Your motion for the latter cannot be represented
by a single vector. As a result, the size of the total distance is different
from that of the displacement.

Charlotte, NC

Columbia, SC

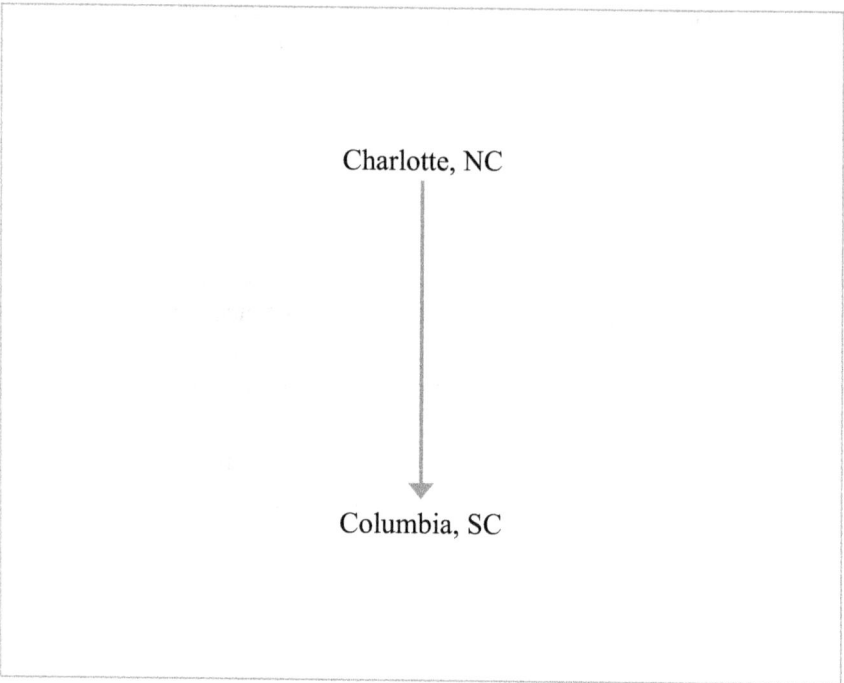

Figure 2.2: Let us go over the case one more time. The mechanical motion could be represented as a vector, with a starting position at Charlotte and an ending position at Columbia. We do have two positions associated with a single vector: the starting and the ending positions. Why? That is what mechanical motion is about, and that is all that matters. A motion in physics can be represented as a change in position; there was an object at a point at a certain time and the object happened to be positioned at another point after a certain amount of time elapsed. For instance, I was in Charlotte at 3:00 a.m. today, and I moved a certain distance and ended up in Columbia at 5:00 a.m. There was an entity at a certain point, and then it moved to another point. What do we need to know to describe the motion? We need to know where I was and what time I was there. All we need to know is the position and the time, which defines a motion in classical mechanics. Now, we may need to know "who" moved from one place to another; thus, the "mass" of the object may need to be specified, but that is not what we cover in kinematics. Just remember that a motion can be represented by a change in position and time. For now, that is all we need to know. We will study more later. It is going to be interesting.

Remember: If I were your instructor, I would emphasize the importance of understanding the essence of vectors over and over, particularly the reason why vectors cannot be added or subtracted as we do with scalars. Lots of students who do not focus on this part and move on have a hard time when trying to understand mechanical motion and Newton's laws later. So, you need to firmly grasp the difference between vectors and scalars to understand the more advanced materials that will be covered in the remaining part of this book.

Problems:

Think about why it is important to understand vectors in mechanics. Figure 2.2 might help you to understand the question more clearly. Write your thoughts in a paragraph.

Describe the difference between vector and scalar and describe why it is not possible to represent someone going from Charlotte to Columbia using only a scalar but a vector. Find a quantity that can be represented by a scalar.

Draw a vector that represents someone going directly from Charlotte to Columbia. Then, on your drawing, draw vectors representing someone going from Charlotte to Atlanta and then Atlanta to Charlotte. Describe the difference between the two routes.

Draw a vector that represents someone going directly from Charlotte to Columbia, and then draw a vector that represents going directly from Columbia to Charlotte. Describe how they are different in terms of vectors.

You have two vectors, vector 1, whose magnitude is 10 meters with an angle of 40 degrees with respect to the horizontal direction, and vector 2, whose magnitude is 20 meters with an angle of 70 degrees with respect to the horizontal direction. Calculate the magnitude and the angle with respect to the horizontal direction of the sum of the two. You need to calculate horizontal and vertical components of each vector, add ones in the same direction, and then calculate the total. Then subtract vector 2 from vector 1.

Day 3
Fundamental quantities

> "Think of minimum quantities we need
> to know when describing a motion. What
> fundamental quantities do we need to
> know when describing motions associ-
> ated with a single object?"

In the previous lesson, we learned that we need to have vectors to describe motion, particularly in classical mechanics. Today, we are going to discuss "with what" quantities we will be able to describe the motions. That may sound too basic, but it is important to understand why we play with such and such quantities when we study motions in classical mechanics in the future.

By the way, we are going to utilize going from Charlotte to Columbia over and over throughout this chapter. You are going to be so familiar with the story in the end. You might end up sick and tired of it because you are going to see a few more lessons where that case is illustrated.

Let us think about motion again. Can you guess "what physical quantities" we need to know when describing a motion? Let us not think too hard on this one but think about the minimum number of quantities that we need to know when describing motions associated with an object. What do we at least need to know? What do we study in this book?

We do at least need to know a departing place and a destination, just like my moving from Charlotte to Columbia in the previous lesson. So, we are good with the fact that we need to know information regarding "where." Again, we need to know the "where" information. That is one of the pieces that we need to know, as far as the change in position goes. By the way, the information regarding "where" allows us to calculate displacement, which will be covered later, so be ready. Now, is there any other information that we need to know when describing motion?

We need to know information regarding "when"; we need to know something about "how much time" to practically describe motions in classical mechanics. Basically, what we discussed in the previous lesson was about comparing how long my trip takes if I take two flights instead of driving directly down to Columbia. So, in the end, it was about comparing "how long" a motion takes. We had to have the time information. In other words, to fully describe a motion in classical mechanics, we need to know at least the location, or I would say, the position associated with the departing place and destination. On top of that, we also need to know how long it takes to get to the destination. The former is what defines motion in physics and the latter, at least in classical mechanics, is needed for an object to change its position. We are not talking about space travel at this moment; thus, we need the latter to realize motion.

> We need information regarding the
> change in position and time to fully
> describe a motion or motions in classical
> mechanics. That is the minimum informa-
> tion that we need.

Now, here comes a simple but a rather important and interesting question: how do we describe them? How do we describe the information regarding "where" and "when"? You may think that it is just about knowing the name of the two cities, one for the departing place and another for the destination. Well, that is a good starting point, but if you think about it a bit, when we name two cities we already assume something. What is that? Yes, from the name of the cities, we can guess the distance between the two cities.

Here comes another important question: how do we describe the distance? In other words, do we need some sort of standardized way to describe the change in position, and if so, what do we have to begin with?

Answer: Yes, we need some sort of "agreement" among people when describing the information regarding "where" and "when," and that is where "units" come in physics. We do need to have the units associated with the minimum quantities. We are not just playing with numbers. We will play with numbers with units at the end. It is for our convenience, but we do need them in order to be "consistent."

We can write the distance in miles, for instance. It is one of the most common units when measuring distance. The distance between Charlotte and Columbia is about 100 miles. It may sound simple but writing the distance as "100 miles" is quite an important thing. What do I mean by that? Question: can you write the distance with a just single number? For instance, can you write the distance as "100" or "1,000" or any other number?

Answer: No, we cannot. Why? Because distance is a "physical" quantity, so we need a "physical unit" to describe a physical quantity such as distance. We need to have something that can be commonly compared with when describing a mechanical motion. Have a look at Figure 3.1.

We need to have something that can be commonly compared with when describing a mechanical motion. Physics is all about comparing something that gets changed.

We don't just discuss numbers here in classical mechanics. Why? People could come up with their own units in their minds and express the numbers only. Think about what would happen in that situation. That is where the importance of using units comes in. We are here to discuss physics, where quantities are associated with physical motion. The "100" is just a number, but "100 miles" is a physical quantity. You cannot simply describe a distance using a purely mathematical number such as "100" or "1,000." When we attach a physical unit such as "miles" to a number, the result is going to attain its meaning as a physical quantity. Why? Again, we all can agree with "how long" a mile is. That is why. We can be consistent with that.

Therefore, when writing the distance as "100 miles," the number "100" is just an expression for a scaling factor associated with the unit "mile". We simply multiply them together. So, this way, the number 100 gains its meaning in physics.

The number "100" is a scaling number to the physical unit "mile," and there it gains its meaning.

Now, here comes another important question: how do we know how far 1 mile is going to be? Answer: We just made it up. Many years ago, people around the world gathered to think about it. They wanted to standardize the distance unit, and they decided that "mile" would be a unit for distance. A mile is an arbitrary unit in length, and we agree to assign a "certain" length to the unit "mile." Again, we made the unit to be a certain length that we know as "1 mile" in physical space. So, for instance, the length we know as an inch or centimeter on a ruler is just what people, at some point in time, had decided to be that length. We agree upon standardized units and utilize them.

Shifting gear to another quantity, what about time? Just like distance, we have a fundamental unit for time, and we set a certain lapse of time to be a second. Why do we need time? We need time for when there is a change in physical units. If there were no time, then there would be no change in position. We are not talking about time travel or space travel here, although I think that is a highly interesting topic that can be covered later in another book.

In short, we certainly need, at a minimum, two fundamental and physical quantities to describe a motion in space: distance and time. The former is to indicate the change in position, or location in physical space, and the latter is to indicate how long the motion takes to move from the initial to the final position. The latter is needed to realize the former in the classical regime.

Here comes another bonus point: when fully describing a motion, with respect to something, we need to describe "how massive" the entity was that moved from one place to another. It is what is called "mass" in physics. We are going to go over what mass is later when we study dynamics, where we have more than one body in a system. For now, remember that we need to know distance and time when describing a motion of a single object in space, and we need physical units to describe the two quantities in physics. It cannot be numbers only, but those numbers multiplied by the physical units. That is how quantities gain their physical meanings—thus, why they're important when studying classical mechanics.

Person 1: I moved from
Charlotte to Atlanta in
0.083 days.

Person 2: I moved from
Charlotte to Atlanta in
100 minutes.

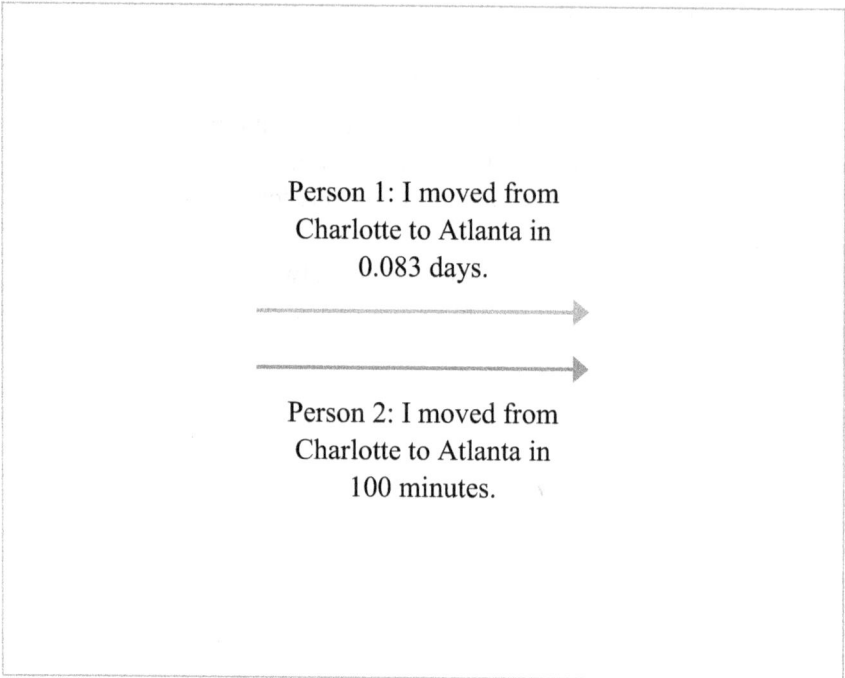

Figure 3.1: This illustrates why it is important for us to describe physical quantities in the same unit when comparing them. Two people traveled from Charlotte to Atlanta. They departed Charlotte at the same time. One of them said the total time taken was about 0.083 days, and the other person said that he or she took about 100 minutes. If they say the total times in two different physical units, we cannot easily compare the quantity of time. One of them should convert the unit associated with the quantity for the other. Either person 1 needs to say their total time in the unit of minutes, or person 2 needs to say their time in the unit of days. Or they both need to convert their units, for instance, to hours, a single unit. The day unit could be converted to minutes; 0.083 days of time is about 120 minutes. Then we can compare the two. Person 1 took 120 minutes to go from Atlanta to Columbia, and person 2 took 100 minutes. Then, when we compare the numbers 120 to 100, we know that person 2 took less time than person 1.

Remember: This is important to understand the difference between mathematics and physics in general. In classical mechanics, we try to understand motion in terms of a change in position and time associated with entities. In doing so, we are not only playing with numbers but also with the associated physical units. We deal with physical quantities. Imagine you only had numbers to play with. Would we be able to describe motion in classical mechanics or not?

Problems:

What we understand and define as "mass" in classical mechanics is not as important as quantity when studying the motion of a single object. In other words, if there is an object in a system and it happened to be the only object in the system, then the mass associated with the object is not important for understanding the motion of the object. Think about why that is so and describe a reason or reasons for that in a paragraph.

What is "time" in physics? Find one or two definitions and think about why introducing time is something that is needed when studying mechanical motion in classical mechanics.

Day 4
Displacement

"Displacement is the shortest distance
between two positions in space and is
represented by a vector, whereas distance is a literal distance an object takes
to move from a position to another as
a scalar."

You know what? Going back to the story of my flights, the agent in the
Charlotte airport as introduced in the first lesson knew not only the difference between the scalar and the vector but also the difference between
displacement and distance. The agent knew everything.

Going back to the different routes I could take, one where I could go
directly to Columbia and another where I was taking two flights to go
there: the former was going directly from Charlotte to Columbia where
I took a shortest route from the departing place to the destination, and
the latter was going from Charlotte to Atlanta and then from Atlanta to
Columbia. Why am I bringing this up again? I am doing so since it is a
good example to illustrate not only the difference between scalar and vector but also the difference between distance and displacement. The customer agent, knowing the displacement for the two routes was the same,
was wondering if I should take the route where the distance is shorter so
that I could save time in going to my destination. What matters in the end
is my arriving at my destination, and that was all the agent cared about.

By definition, distance in physics is the literal distance that an object
or objects move in space. So, in the former option, the distance that
I would have moved is 100 miles total, and that distance happens to be
the shortest distance from Charlotte to Columbia. I moved 100 miles, so
the distance is 100 miles. The distance and the displacement were the
same for the former option. I took the shortest route to the destination.
Figure 4.1 might help you understand the point more clearly.

But the case was a bit different for the latter. I would have moved
400 miles from Charlotte to Atlanta and then 300 miles from Atlanta to
Columbia. So, what would my total distance have been? Here the total
distance would have been 700 miles. You simply add them together.
How about displacement? Is it the same as the distance? No, it is not.

As I mentioned earlier, what matters when calculating displacement is where the starting and the ending positions are and how to calculate the shortest distance between the two positions. The shortest in this case was 100 miles, which represents the size of the line that connects Charlotte to Columbia. In other words, displacement is a function of the start and end positions. You can think of my moving from Charlotte to Atlanta as a vector and from Atlanta to Columbia as another vector. Adding them together, we are going to end up getting the vector that represents my directly going from Charlotte to Columbia. I would have moved literally 700 miles, but the displacement was only 100 miles.

Why am I bringing up the same case here? I am doing so to point out that the displacement in both cases was the same, but the distance was not. Think for a moment. The departing place is Charlotte, and the destination is Columbia for both. It is just that there was a stoppage in the latter; a different path was taken, but the "effective" distance, which is the displacement, of my moving were the same in both situations. The difference between the two positions, in terms of representing it as a vector, was the same. By the way, we care more about displacement than distance when studying classical mechanics, and you are going to see why when studying dynamics.

> Displacement is a shortest distance
> between two positions, and such could be
> represented as a straight line taking the
> two positions as end points.

Now, we can make the situation a bit more complicated. We can think of another route; again, the departing place and the destination are going to be same but, for instance, I can make two stops; I can move from Charlotte to Atlanta, Atlanta to Chicago, and then Chicago to Columbia. On the same token, you can think of a more complicated route if you want to. In the end, there are an infinite number of routes that I could take, but no matter which route I take, the "effective" distance of my moving from the departing place to the destination is going to be same. Each different route has a different "total" distance, the distance I would have to move in total to get to the destination from the departing place. That is what displacement is all about, and this point is going to be the basis when studying special theory of relativity later, if you are interested. There just happen to be some extra constraints that we need to think about.

In short, displacement is the same as the "effective" distance. It represents the shortest distance from one place to another. When the departing place is Charlotte and the destination is Columbia, then the displacement, or the shortest distance, is always 100 miles, no matter which route is taken. This will be same as the total distance if the shortest route is taken but will not be the same if a different route is taken.

Why is this important? Because many kinematical quantities that you will deal with in physics class are based on displacement, not distance. Furthermore, understanding the difference between the two is going to help you better understand the fundamentals of vectors too. Just remember that displacement is the shortest distance between two points in space, but distance is literally the total distance that an object takes.

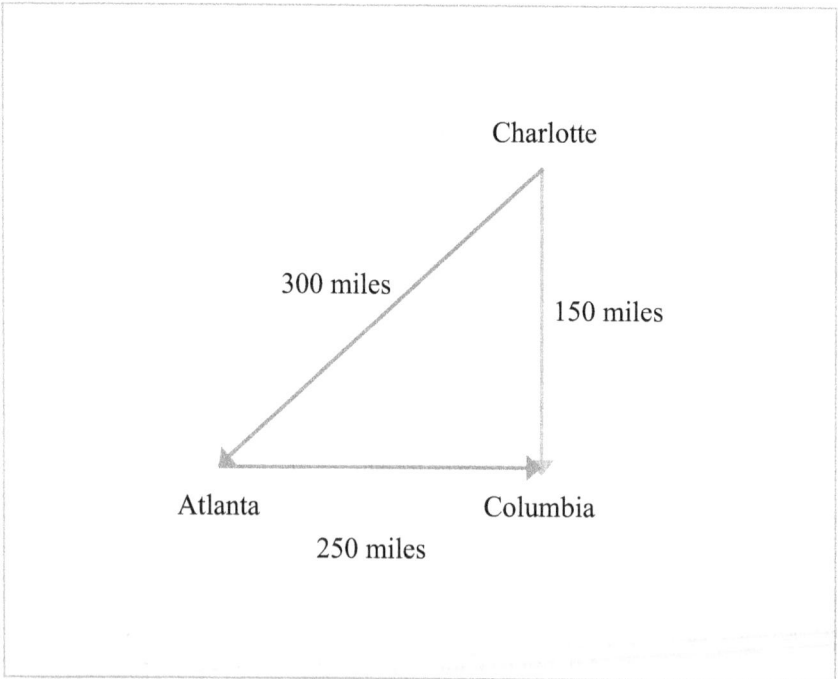

Figure 4.1: This shows the difference between the displacement and the distance. Displacement is the shortest distance between two positions, and distance is the distance that an object moves. When going from Charlotte to Atlanta and then from Atlanta to Columbia, the total distance was 550 miles since it is a simple sum of 300 and 250 miles. But the displacement in this case is only 150 miles. Displacement is all about the shortest distance, so whether I went to Atlanta first or not, it does not matter. As long as the starting position was Charlotte and the ending position was Columbia, then the displacement is going to be 150 miles.

Remember: Understanding differences between displacement and distance in classical mechanics is another chance for you to grasp the difference between vectors and scalars before moving on to more advanced topics. If you do understand why displacement is different from distance, you can move on and try to understand derived physical quantities such as velocity and acceleration.

Problems:

Describe in a paragraph the difference between displacement and total distance using the example of my moving from Augusta to Charlotte and then Charlotte to Columbia. Google the distance from Augusta to Charlotte and from Charlotte to Columbia and find out the displacement.

An instructor had a visit to Charlotte. On his way back, he had two flights, one from Charlotte to Atlanta and another from Atlanta to Columbia. Calculate the total displacement of his moving from Charlotte to Columbia. Calculate the total distance he moved in the two flights. Assume that the distance from Charlotte to Columbia is 100 miles, from Charlotte to Atlanta is 300 miles, and from Atlanta to Columbia is 200 miles. In physics, distance is different from displacement.

Imagine that you had a trip from New York to Chicago and then from Chicago to Washington. Can you define a "direction" that is associated with your entire trip? If you can, describe your reason. Do the same if you do not think so.

We have an advantage of our using displacement over distance in terms of a reference that we need to set when describing a motion. What advantage is that?

Day 5
Velocity

"Velocity is a quantity that describes
displacement as a function of time. You
have displacement and time. You just take
the ratio of the two, and you do this for
your convenience."

Again, believe it or not, the customer agent in the Charlotte airport knew
more than just the displacement. She knew about velocity in physics
too. The agent knew that it was going to be faster when moving directly
from Charlotte to Columbia instead of taking two consecutive flights.
Figure 5.1 might help you to understand the point better. Which one was
faster? Wait a second. How does she know all that?

   As described earlier, position and time are the two fundamental quantities that we may need to use when describing the mechanical motion
associated with an object. We probably need one more fundamental quantity called mass when describing kinematics of more than one object, but
when we focus on a single object in a system, the two quantities, position and time, are what we need to heavily deal with here in kinematics.
Question: do we always need to deal with the position and time information separately, or is there some other way for us to use them more
wisely, for our convenience? In other words, do we need to deal with the
change in position in that time and have that be the end of the story for
our lessons? Answer: We have something that we can "come up with"
based on the position and the time information, and that quantity happens
to be called "velocity," a highly important quantity when studying kinematics. Velocity is defined as the change in position, or displacement,
over time, and it indicates how fast an object is being displaced during a
certain time. In other words, you take the displacement and divide it by
the time it takes to realize the displacement. Why do we do that? Again,
we do that for our own convenience.

Velocity is the ratio of the change in position associated with an object
over time. In other words, you take the change in position and divide it
by time, and you end up with velocity.

Let us go back to the case of my moving from Charlotte to Columbia but using displacement only. We assume that the shortest distance between the two cities is 100 miles. Now, let us think about the time it takes to go from Charlotte to Columbia; it takes about 10 minutes when taking an airplane, but it takes about 100 minutes when we drive to the city of Columbia. What is my point here? The same displacement does not always lead us to the same motion. Time could be different.

We want to be specific when it comes to describing motions.

We are not comparing the case of taking two different displacements but taking the same routes. Displacement is going to be the same, but the time it takes to realize the displacement is going to be different.

In practice, velocity indicates how "fast" or "slow" an object moves from one place to another. For instance, in my moving from Charlotte to Columbia in about 10 minutes, the velocity is going to be 100 miles divide by 10 minutes, so it is 10 miles per minute. On the other hand, when flying to my hometown in an airplane, the velocity is going to be 1 mile per minute. That is because 100 miles per 100 minutes is same as 1 mile per minute.

Now, coming back to the definition itself, you may ask the following question: why is this quantity, "velocity," so important in physics? There could be many different reasons, but if you need to choose one, then it would be the following: it is easy to get an immediate sense of "how fast" an object moves. If the size of the displacement is the same, we want to analyze the motion with further details. By just using velocity, I could immediately see that the airplane would carry me to the destination faster than a car from Charlotte to Columbia. Again, it is for our convenience. The velocity of my moving was 10 miles per minute when using an airplane and was 1 mile per minute when using a car. Just comparing the velocities enables me to see which one is faster. There is no need to mention the distance and the time that something takes to be displaced every single time, but just the velocity, and there we can easily see that it is the airplane that moves with a higher velocity.

Here comes another important reason: velocity can be "compared" no matter where the initial and the final position of an object are; for instance, I moved from Charlotte to Columbia in 100 minutes, and someone takes about 100 minutes to move from Virginia to Charlotte. The size of the displacement from Charlotte to Columbia and from Virginia to

Charlotte is about 120 miles, so, assuming that I am going to use a car when moving and the direction of the two vectors are parallel to each other, the velocity is going to be the same, although the departing place and the destination can be different. That means that we can even compare my moving from place 1 to place 2 with my moving from place 3 to place 4. Velocity is a quantity that allows us to do so, no matter where the start position and the end position are located. Figure 5.2 might help you to have a better understanding of the point there.

In short, velocity is a useful quantity that is derived from position and time. We can use that to quantify how fast an object moves. By the way, there is a bit more to velocity in physics, and we are going to study them later in the dynamics part.

Charlotte, NC
June 20, 2018,
10:00 am

Chicago, IL
May 10, 2018,
8:00 am

Columbia, SC
June 20, 2018,
11:30 am

Atlanta, GA
May 11, 2018,
11:00 am

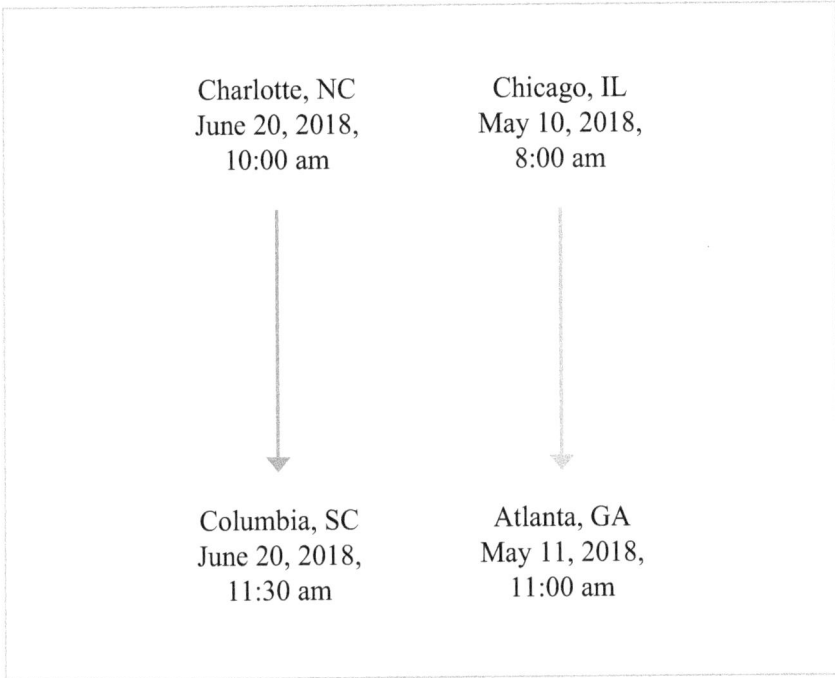

Figure 5.1: This illustrates the importance of having velocity as a kine-matical quantity in mechanics. Position and time information are two basic pieces of information that we can get, but, in the case shown in the figure, if we just compare their change in position and the time, it will be hard to compare the motion of the two. From Charlotte to Columbia, the distance is about 120 miles, and it took about 90 minutes. From Chicago to Atlanta, the distance is about 1,200 miles, and you need about three hours. By just comparing the given numbers, it is hard to tell in which case I was moving faster. However, by calculating velocity, the rate of change of position with respect to time, we know that it is 80 miles per hour for the former and 400 miles per hour for the latter. So, we know the former was slower than the latter. With that, we can simply compare the velocity, and it is easy to get some sense of their speed.

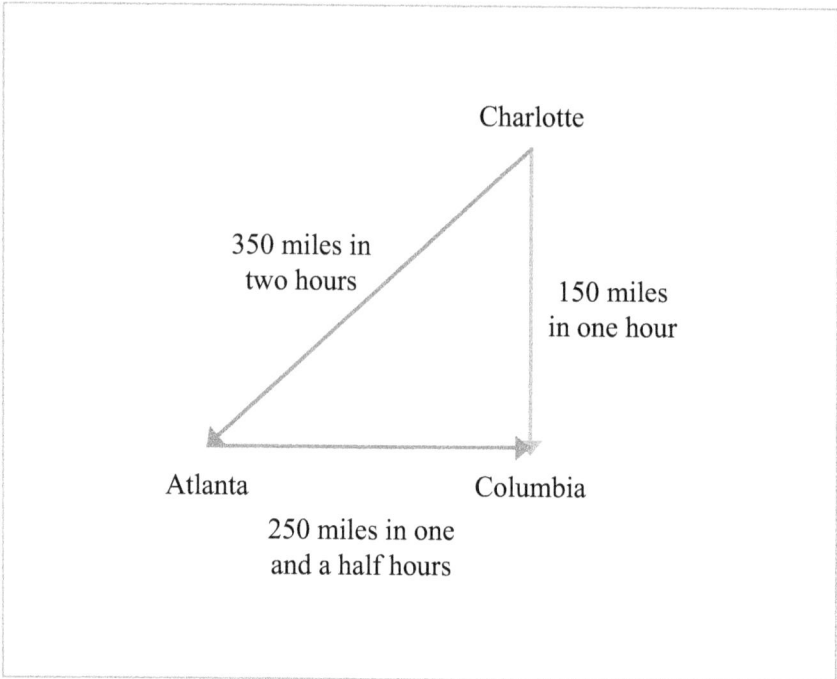

Figure 5.2: This illustrates the difference between velocity and speed. Remember that velocity is a vector, and speed is a scalar. When going from Charlotte to Columbia directly, the velocity and the speed are going to be the same. The velocity is going to be 150 miles per hour, and the speed is going to be the same. However, for the case of going from Charlotte to Atlanta and then from Atlanta to Columbia, the velocity is going to be the total displacement over the time it takes, which is 150 miles of displacement over 3.5 hours, since that is the total time that was taken. So, we end up with 42.85 miles per hour, whereas the speed is going to be the total distance over time, so it is 600 miles of distance over 3.5 hours, and so the speed is about 171.42 miles per hour. In other words, when the displacement and the distance are different, the velocity is going to be different from the speed. Do you see where the difference between velocity and speed is coming from in this case?

Remember: Once you understand the basic notation associated with velocity, you are ready to move on and try to understand what acceleration is about in classical mechanics. However, keep in mind that the physical aspect associated with velocity and acceleration is different. The mathematical expression for the two might look similar to each other though. We are going to go over them later when we study dynamics.

Problems:

The definition of velocity is displacement over time. That means that you take the displacement and divide it by the time it takes for the displacement to be realized. Taking that into account, calculate the velocity of my moving from Charlotte to Atlanta in two hours and then moving from Atlanta to Columbia in three and a half hours. Assume that the size of the displacement from Charlotte to Atlanta is 300 miles and from Atlanta to Columbia is 200 miles.

Google the definition of speed and describe in a paragraph the difference between speed and velocity.

Assume that raindrops fall in a vertical direction at 100 miles per hour and you are running in a horizontal direction at 150 miles per hour. Calculate the velocity of the raindrop hitting your face. This is about understanding that a physical quantity or quantities are relative depending on where we set a reference, which is something that was covered in the very first lesson in this book. The speed of the raindrops is 100 miles per hour with respect to the ground, but it is not going to be the case anymore when it is measured with respect to someone or something that runs in a horizontal direction.

Imagine that you push something that was at rest, thus it gains a certain velocity. Then, somehow, for some reason, all the entities in our universe, including yourself but except the object, disappeared. In other words, the object is the only object present in the universe. Describe what will happen with the velocity and all other physical quantities associated with it. Hint: you may want to read the lesson on Newton's first law in this book. It is in the dynamics part.

Day 6
Acceleration

"Acceleration is the change in velocity
as a function of time. It is going to be a
basis for our defining force. This is where
kinematics meets dynamics in classical
mechanics. We always need to do this
when we have more than one object
in a system."

Question: imagine that you have a trip to New York from your home-town. Can you drive your car the same velocity the whole time? Have a look at Figure 6.1.

Once you understand what velocity is about, it is time to study what acceleration is. Just like the way we define velocity as the change of something over time, acceleration is the rate of change of velocity as a function of time. That means that the size of the acceleration tells us how fast velocity changes as a function of time. Once you grasp the essence of the velocity, you can think of the acceleration in the same way.

Acceleration indicates how fast velocity changes.

It is as simple as that, although we have other reasons for studying and understanding acceleration in classical mechanics, particularly regarding Newton's laws, something that you are going to study later. Anyway, we are going to go over that later in the dynamics chapter, so be ready.

Let us go back to the case of moving from Charlotte to Columbia again, and this time let us assume that I was driving a car. It is easy to see that you will move slowly until you get on the highway, and then you will move fast while you are on the highway, and then you will go slow again once you exit from the highway. The point here is that the velocity of movement going from Charlotte to Columbia is not going to be same the whole way. If it were, you could get in trouble. Imagine driving a car 80 miles per hour on a local street. You could end up in a jail.

Point: velocity often changes in the moving motion of an object or objects, and we want to quantify the change in velocity as a function of time, and that is where acceleration is introduced as a highly useful quantity. It measures the rate of change of velocity as a function of time. In other words, velocity is the change of position with respect to time, and acceleration is the change of velocity with respect to time. It is a change of a change.

> Acceleration is the change of veloc-
> ity over time or the rate of change of
> displacement over time.

If you studied the velocity part and now the acceleration part and see the similarities between the two, then you might ask why we do not introduce more quantities to describe the change of acceleration with respect to time and so on for the following quantities. We can certainly do so, and if you want to analyze motion where the acceleration does not stay the same, then we need to do so, but in most situations we are going to analyze motions in parts where the acceleration stays as constant, practically speaking. Understanding the essence of kinematics here in classical mechanics, we may not need to go down further into the rate of change of the acceleration.

There is another reason that acceleration gains importance. Acceleration that is associated with an object or a system of objects is directly proportional to a highly useful and dynamic quantity called "force," something that you will study later. The force that an object has can be defined as the mass of the object multiplied by its acceleration, and force happens to be an important quantity to understand dynamics in mechanics, so there the acceleration gains its importance. But you know what? It is a quantity that we are going to play with when studying interactions between objects, not a single object but more than one, so we are going to study another aspect of acceleration later on. It is going to be interesting.

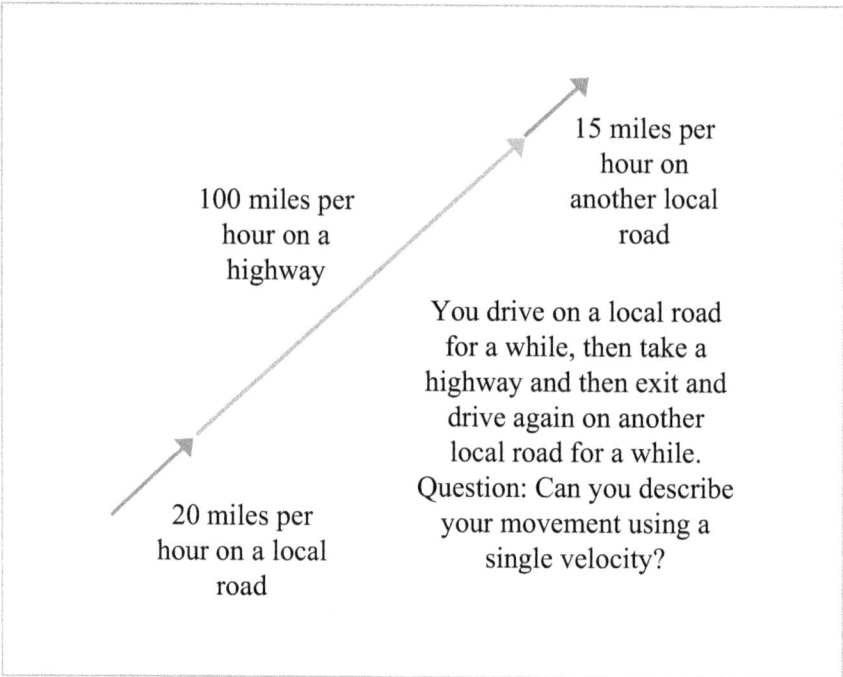

100 miles per
hour on a
highway

15 miles per
hour on
another local
road

20 miles per
hour on a local
road

You drive on a local road
for a while, then take a
highway and then exit and
drive again on another
local road for a while.
Question: Can you describe
your movement using a
single velocity?

Figure 6.1: This illustrates why we need a physical quantity including acceleration in order to describe a more realistic motion. For example, a person drives his or her car from a home to a school. Question: can the person drive 80 miles per hour all the way from their home to the school? Probably not. If they did, as you can imagine, the person could get in trouble. The point is that his or her motion cannot be described by a single number representing the velocity in general. Instead, we may need to have a quantity such as "acceleration," a quantity representing a change in velocity as a function of time elapsed. You can think about what we need to do if the acceleration changes. I will leave that up to you.

Remember: acceleration is the change of velocity as a function of a change in time, but that does not mean that acceleration does not change at all. Acceleration can certainly change with respect to time, for sure. It is just that you might end up assuming that it is not changing as a function of time when working on a certain type of problem.

Problems:

Acceleration is something that can be measured; we can measure the distance or displacement someone moved using rulers or some other devices, and we can measure time that an object or a person takes to move that distance, so we can first calculate the velocity and then acceleration as taking the derivative of the velocity with respect to time. Think of a situation where acceleration stays "almost" the same all the way from the beginning to the end of your measurement and describe how you estimate the magnitude of velocity when acceleration stays the same.

Think about how you can measure the time for a small object to be displaced by 100 cm in the vertical direction.

"We can certainly analyze a case where the size of the acceleration is not going to stay the same. However, even when we do so, we are going to most likely analyze the case in parts where the acceleration stays nearly the same over time." Think about why this is so in most cases when studying classical mechanics. Imagine that we have only two objects in a system that we are interested in and think about what happens. Hint: you might want to come back to this one after spending a few minutes reading about Newton's second law.

Can we ever generate acceleration on our own? Imagine that there is a single object present in our universe and that is the only object. If so, how come? If not, why not?

Day 7
Velocity and acceleration

"Acceleration is change of velocity as
a function of time. In other words, you
need to have a nonzero acceleration in
order to realize a change in state."

Question: velocity and acceleration are related to each other, but how come, mathematically speaking?

If you studied the previous lessons and understood how velocity and acceleration are related to each other, you may jump to the next lesson, but it is going to be helpful to go over the lessons from another perspective.

Believe it or not, we are now speaking the kinematical quantities more in the language of classical mechanics. What does that mean? Let us think about velocity for a moment. You can express the velocity in a few different manners. We have two of them that we generally go with in common; one in terms of displacement and time and another in terms of acceleration and time. The former is just the way that the velocity is defined in one of the previous lessons: velocity defined as a change in position over time. So, the change in position divided by the time is what velocity is, although we need to be a bit cautious in defining "what" velocity we are talking about there. For now, let us just take it as it is and move on.

On the other hand, the latter is another way of defining velocity. Instead of position, we are going to define the velocity in terms of acceleration. In other words, we define velocity as a function of acceleration and time instead of displacement and time; change in velocity can simply be written as acceleration multiplied by time, with an assumption that the acceleration stays the same during the time; the acceleration remains as a constant,

Change in velocity = Acceleration × Time

It requires a few arithmetic operations to show the relations among velocity, acceleration, and time, so if you are interested in this, you may find the mathematical relationships in other references. Here, we just accept that the change in velocity can be written in such a manner and move on.

> Change in velocity is acceleration
> multiplied by time when the acceler-
> ation as a vector does not change as a
> function of time.

It is equivalent to say that the final velocity is the sum of the initial velocity and the acceleration multiplied by the time elapsed. Why is this important? Answer: This is important because we now can calculate the size of the velocity if we know the acceleration and time information. On the same line of thought, acceleration can be calculated if velocity and time are known, and time can be calculated if velocity and acceleration are known. Among the three variables, two are known and one is unknown. You just calculate the size based on the relationship above. Believe it or not, once you realize this, it is just a matter of punching the numerical quantities into your calculator when working on some practice questions.

Understanding the point is going to be important, especially when we study the dynamics part, which is all about understanding Newton's laws. Acceleration has to do with what is called, force, a quantity that needs to be introduced in dynamics and that can be written as mass multiplied by acceleration. Velocity has to do with what is called momentum, a kinematical quantity associated with the mass and the velocity of an object in motion. Given how velocity, acceleration, and time are related to one another, we can relate the two dynamical quantities and describe the motion of an object better and clearer where gravitational interaction is present.

The way acceleration is associated with change in velocity is about associating the dynamics to the kinematics part in classical mechanics, and that is exactly how writing the velocity as a function of acceleration and time gains significance; we need to know the relationship between the velocity and the acceleration first to understand the dynamics better. This might be something to think about later, but for now, just remember that the change in velocity can be written as acceleration multiplied by the time it takes. If velocity and time are known, we can calculate acceleration. If velocity and acceleration are known, we can calculate the time. In short, if two out of three quantities are known, we can calculate the one that is not known.

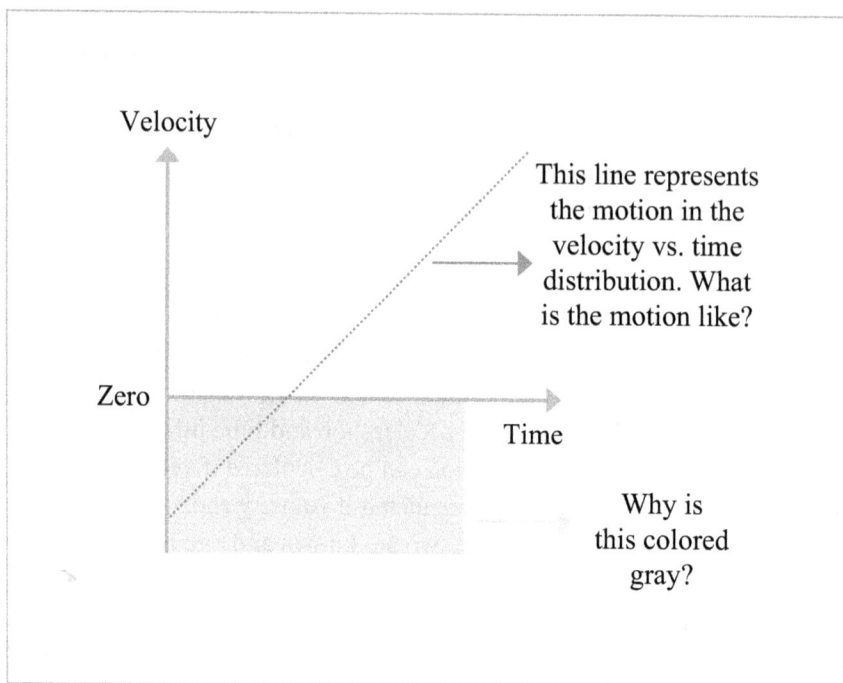

Figure 7.1: This illustrates a motion in velocity vs. time. The vertical line represents the size of the velocity associated with an object moving in a one-dimensional space, and the horizontal one represents the time. Imagine that there is a car moving with a constant rate of acceleration in the dimension. How is the motion associated with the object going to be illustrated in velocity vs. time distribution? How about with a constant velocity? Hint: you may want to think about what "slope" represents in the velocity vs. time distribution.

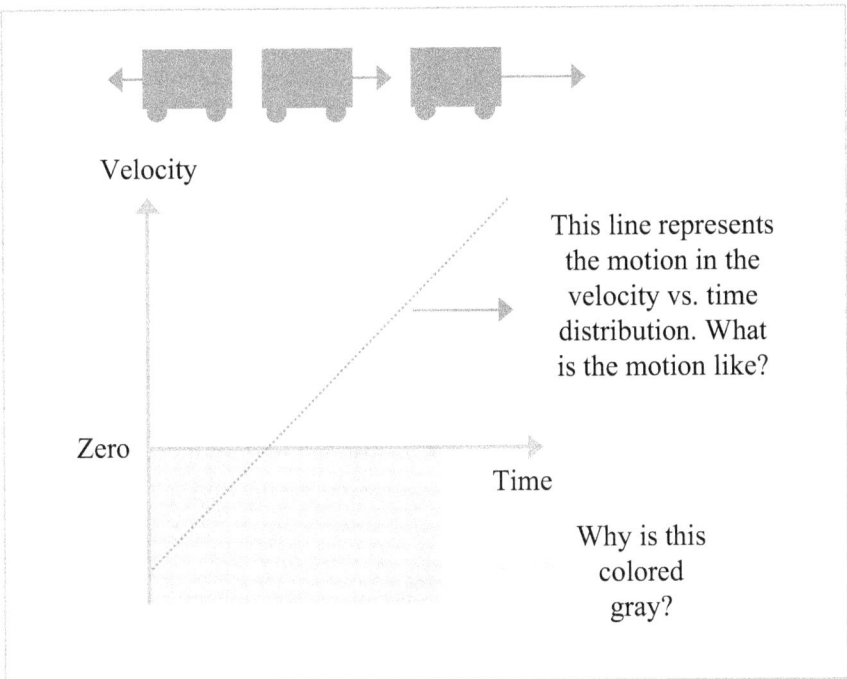

Velocity

This line represents
the motion in the
velocity vs. time
distribution. What
is the motion like?

Zero

Time

Why is this
colored
gray?

Figure 7.2: This is the same as the previous one, but there are three cars
with arrows on the top. Imagine that the motion of the car is represented
by the velocity vs. time distribution. What does it have to do with the
arrows associated with the cars in gray on the top. Why is one of them
going backwards and the other two are going forward?

Problems:

Go over Figure 7.1 and think about how an object moving with a constant rate of acceleration is going to be represented in a velocity vs. time distribution, where the vertical line represents the size of the velocity, and the horizontal line represents the time.

We can certainly write the change in the size of the velocity in terms of acceleration and time "only." However, when we write the relationship in such a manner, we need to make a critical assumption. Think of what the assumption is going to be and think about what happens if we do not assume when writing the change in velocity only in terms of acceleration and time. Hint: the rate of acceleration does not need to be the same all the time.

Go over Figure 7.2. Write down the velocity equation as described in the lesson. Imagine that an object was in motion, initially at 10 miles per hour velocity, moving back and with 100 miles per hour squared acceleration for a total of 20 hours. Calculate the distance and the displacement associated with the object in the 20 hours. Hint: you may want to think about why a portion in the distribution is colored gray and what that is really about.

You accelerated at 100 miles per hour squared for two hours. Calculate the size of the change of your speed.

Imagine that the change in the magnitude of your velocity is 200 miles per hour, during which you accelerated at 50 miles per hour squared. Calculate the time during which you were accelerating.

Imagine that you have two objects in motion with the same magnitude of velocity and acceleration. Would we be able to distinguish them? If so, what physical quantity is needed for us to do so?

Find a definition for inertial frames in physics and describe briefly what that has to do with the acceleration, or to be more specific, accelerating frames.

Day 8
Position and velocity and acceleration

"Change in position can be written as
a function of velocity, acceleration,
and time."

Now, we take the expression that we came up with in the previous lesson and move forward. We are going to add one more layer to the previous expression.

By definition, acceleration is a change in velocity as a function of time elapsed and because of this, velocity can also be written as a function of acceleration and time. That is one thing we learned in the previous lesson. By the same token, velocity is a change in position as a function of time, and, because of that, change in position can be written as a function of velocity and time. Since the velocity is a function of acceleration and time, following the simple chain of logic leads us to:

Changes in position, a displacement,
can be written as a function of velocity, acceleration, and time. We assume
that acceleration stays as a constant
during the time.

Or,

Displacement = Velocity × Time + 0.5 × Acceleration × Time × Time

A reason for our having the scale factor in the second part of the expression has something to do with our understanding of calculus. It can be understood as a way of our defining what we named as average velocity, which is just the average of the initial and the final velocity over time.

In essence, when describing motion where the acceleration is constant, all you need to know is displacement, velocity, acceleration, and time. If one of them is not given as a part of a practice question, you can calculate the size from the equation above. If you have studied calculus before, you can derive the expression above from that in the previous lesson, but if not, you can just take the expression as granted for now.

Most problems that you deal with for a mechanical motion associated with a single body in space can be solved using the expression above, and that was covered in the previous lesson. For our convenience, let us name the expression above as a "position equation" and the one in the previous lesson as a "velocity equation." Using the two, you can calculate the unknown based on that which is known or given. It is going to be a matter of properly identifying the quantities by reading the practice questions.

You can find the mathematical expressions for the position written as a function of velocity, acceleration, and time in many other sources of literature. For that reason, I am not going to go over all the mathematical aspects associated with the position and the velocity equation. I just want to emphasize the following: velocity can be expressed as a function of change in position over time; therefore, the position can also be expressed as a function of velocity and time. On the same line of thought, velocity can be written as a function of acceleration and time.

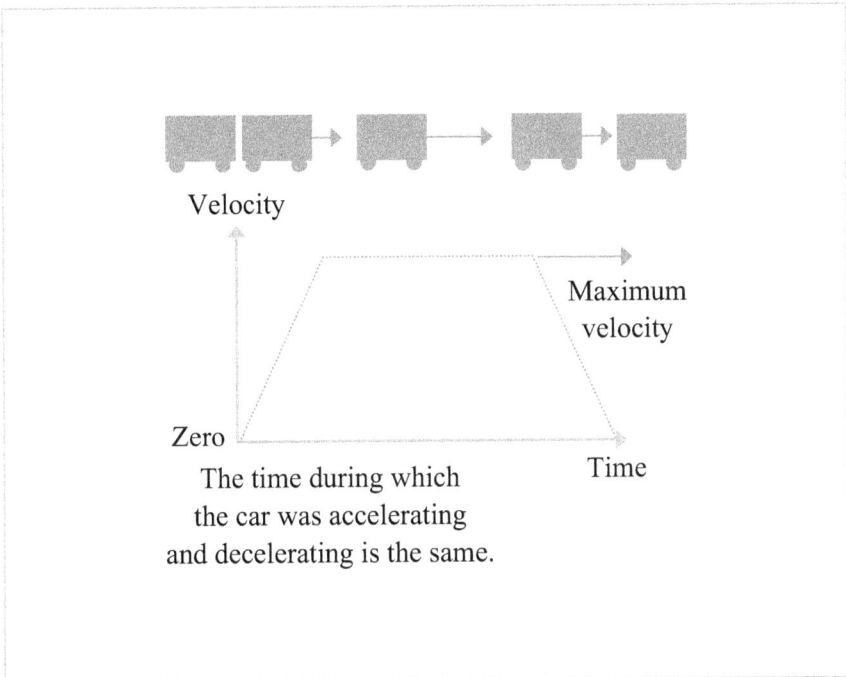

Figure 8.1: Imagine that someone had a trip from Columbia, South Carolina, to Atlanta, Georgia. The trip was about three hours long. The car was accelerating constantly for a certain length of time, moving at a constant velocity of 800 miles per hour for a while, and then decelerating at a constant rate for the same amount of time that the car was accelerating. How long was the car moving with a constant velocity? Assume that the distance between the two cities is 300 miles.

Problems:

Write down the position and the velocity equation.

We can define the change in position, or displacement, as a function of velocity and acceleration and time. However, we need to make a critical assumption: the rate of acceleration remains constant during the time. Describe why we need to make that assumption and describe what happens if the assumption does not hold.

Go over Figure 8.1. Imagine that someone had a trip from Columbia, South Carolina, to Atlanta, Georgia. The trip was about three hours long. The car was accelerating constantly for a certain period of time, moving with a constant velocity of 800 miles per hour for a while, and then decelerating at a constant for the same length of time that the car was accelerating. How long was the car moving with a constant velocity? Assume that the distance between the two cities is 300 miles.

Based on reading the previous lesson and this lesson, write down the velocity and the position equation. Derive the position equation from the velocity equation with an assumption that the change in position can be written as average velocity multiplied by time elapsed. You may assume the other as needed.

Imagine that someone is in motion, with 10 miles per hour squared as the acceleration and 40 miles per hour as the initial velocity, and that the person is in motion for 30 hours. Calculate the size of the total displacement associated with the person. Calculate the same when the sign associated with the initial velocity is changed.

Can we write change in velocity as a function of change in position and time? Think about it and find the right equation from other sources in the literature.

Day 9
Orthogonality

"What is happening in the horizontal direction has nothing at all to do
with what is happening in the vertical
direction."

Question: imagine that you are running on a track as hard as you can. All of a sudden, a strong wind blows up from the side. While the wind blows, does your velocity in the direction parallel to the track change or not?

This lesson may help you understand some of the practical aspects associated with vectors in physics better, so let us go over the basics here first. Believe it or not, understanding this lesson is going to be highly important when you study more complicated cases that deal with projectile motion. It is going to be very important when you also study rotational motion later. You might want to think about the question at the top for a moment before moving on. What does that have to do with the orthogonality in physics?

Imagine that you had a trip to New York from Charlotte on a lovely day. You were driving your car 80 miles per hour, heading straight north. In other words, people watching your car while standing still on the side of the highway are going to "think" and measure the velocity of your car to be 80 miles per hour. Hint: think about the very first lesson we had in this book. Remember that we need to have a reference when measuring some physical quantities from the very beginning. Figure 9.1 illustrates the case to begin with.

Now, here comes an interesting case: say, all of a sudden, a strong wind blows from east to west with a 60-mile-per-hour velocity while you are driving your car as it used to be. So, it is going to be a wind blowing up from the side, and the velocity is huge, so it is something that you cannot simply ignore. Question: can you guess the velocity associated with your car with respect to someone standing still on the ground watching you drive? Think about the very first lesson we studied again. What will the velocity be?

If you have studied the vector lesson and have understood the main point, calculating the answer to the question is not going to be too difficult for you. Answer: It is going to be 100 miles per hour. Why? You simply square 80 and then square 60 and then add them together. Then you take a square root of it. What is this about? Yes, it has to do with the Pythagorean theorem, and you end up with 100 miles per hour as the final velocity. The components of velocity are "orthogonal" to each other, so you have the answer by calculating the size of sum of the two vectors. Have a look at Figure 9.2.

Now, here comes one more interesting point, and this one is important: imagine that someone is moving along with the wind somehow. Yes, it might be a bit weird to imagine the situation but think of someone moving "with" the wind. In other words, someone is moving 60 miles per hour from west to east. The velocity associated with the person is the same as that of the wind. Question: what is going to be the velocity of your car with respect to the person? Answer: It is going to be 80 miles per hour. We are back to 80 miles per hour. Why? What is happening in the vertical direction is independent to what is happening in the horizontal direction.

What is happening in the vertical direction is independent to what is happening in the horizontal direction.

Figure 9.3 may help you understand the main point better. Let us think of another case, and this might be a case that is easier to deal with. What happened to the velocity associated with the flight?

Another way to think of the case is the following: imagine that the wind does not blow up anymore. Out of a sudden, the wind is gone. Now, can you imagine what the velocity associated with your car is with respect to a person standing still on the ground? Yes, you do not need to do many calculations, and it is a bit easier to imagine that the velocity is going to be back to 80 miles per hour, the velocity of the car before the wind was blowing. The engine in your car is letting you move 80 miles per hour on the ground and that is what that is. The engine was providing that much velocity, and now you are back to the first velocity.

What is the main point here? As long as the "driving force" is provided to a car, it is going to keep moving 80 miles per hour in the vertical direction, no matter what happens in the horizontal direction. What happens in the horizontal direction is not going to impact what happens in the vertical direction. They may share something in common though. The velocity of your car associated in the vertical direction did not change and is not going to change in the future, even with the wind blowing from the side. The wind that happens to blow from the side adds up to the apparent velocity of the car. The car appears to move 100 miles per hour; the "apparent" velocity of the car with respect to the ground surface happened to be 100 miles per hour. It just appeared to be. Keep in mind that the velocity in the vertical direction was not changed before or after. It was there. It was the same as long as the driving force provided by the engine of your car was the same. Do you see the point?

That is what the orthogonality relationship is in essence. By the way, understanding the point is going to be important when you study projectile motion, a motion in two- or higher dimensional space. You are going to play with quantities that are relevant for the motion in a vertical direction and utilize the information in order to quantify the motion in the horizontal direction or vice versa. It is just a matter of identifying something that can be shared in common and something that cannot be shared, such as what we illustrated here in this lesson.

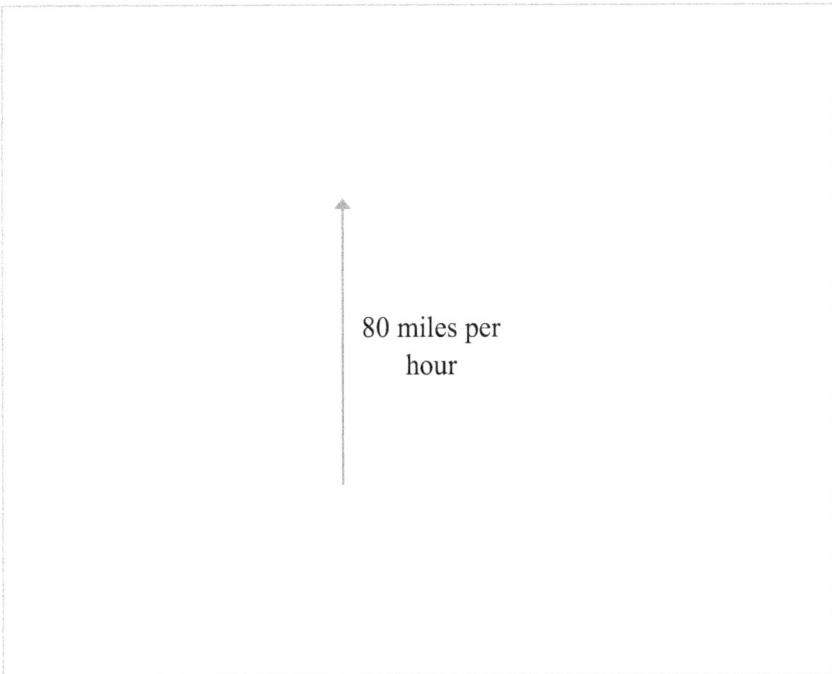

80 miles per
hour

Figure 9.1: You are driving a car heading from south to north. The velocity associated with your car is 80 miles per hour in the vertical direction. Imagine that there is nothing in the universe but you and your car and you are driving. Hint: you may want to read a lesson on Newton's first law briefly.

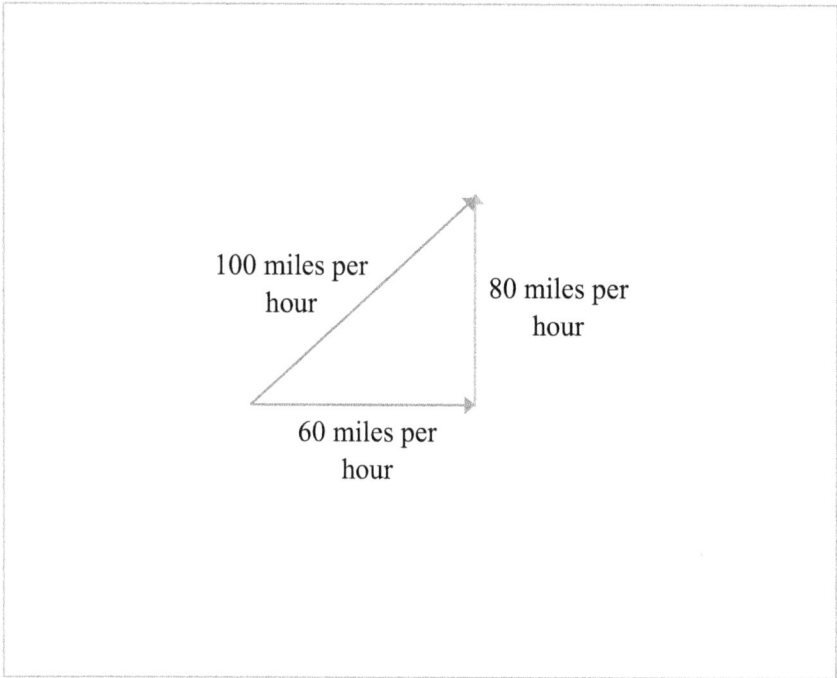

Figure 9.2: Now, we have a small issue: imagine that a strong wind begins to blow from west to east, and the velocity associated with the wind is not something that we can ignore. It is about 60 miles per hour. Here comes an interesting point: imagine that someone is standing still on the ground and he or she is going to measure the velocity of your car. Yes, the measured velocity is 100 miles per hour instead of 80 miles per hour. Remember the very first lesson in this book. Everything is relative. It depends on where we set a reference point. If the reference is some-one standing still on the ground, the velocity of the car is going to be 100 miles per hour.

100 miles per
hour

60 miles per
hour wind

60 miles per
hour someone

In the end, we
are going to
end up with 80
miles per hour.

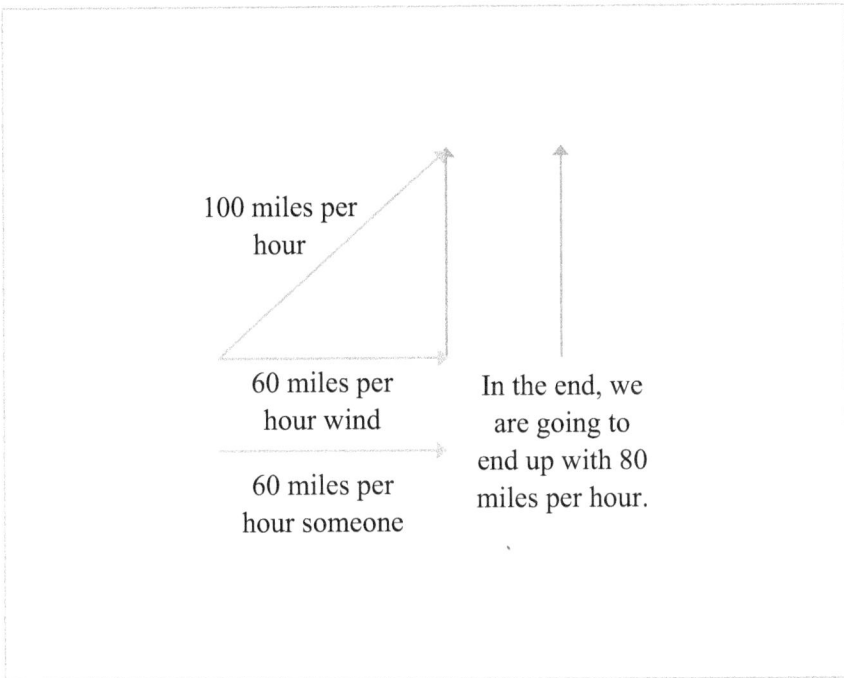

Figure 9.3: However, the situation is going to be slightly different for someone in motion. Imagine that someone is moving along the wind that blows from the side. In other words, the velocity of the person is the same as the velocity of the wind, 60 miles per hour from west to east. Question: what then is the relative velocity of someone with respect to the wind? Yes, it is going to be zero. You add the wind vector represented by a light gray arrow on the top in the left side of the figure and the one on the bottom, and you end up with none. In the end, you are going to end up with 80 miles per hour as the final velocity of your car with respect to someone moving the same velocity as the wind. In other words, with respect to someone moving along the wind, he or she is going to think you are moving 80 miles per hour. Point: what is happening in the vertical direction is not changed by what is happening in the horizontal direction. As long as the size of the driving force is leading your car to 80 miles per hour, that is going to be the same as the 80 miles per hour, no matter what happens in the horizontal direction. If the reference is set to be the same as that of the wind, we can simply ignore the wind when estimating the velocity.

Remember: understanding the differences between vector and scalar is important. Likewise, it is also very important to understand what the orthogonality relationship is, especially when you are going to study a projectile motion, a motion in two- or higher dimensional space. If you do not understand what the relationship is about, you are going to have a hard time firmly grasping how we end up with various physical quantities in the projectile motions in higher dimensional space. For that reason, I highly encourage you to ensure that you understand what the orthogonality relationship is about, especially in the Cartesian coordinate system, before moving on to the topics in dynamics, including Newton's laws of motion.

Problems:

Imagine you are driving a car going 80 miles per hour on a highway and a truck driver is driving his or her truck on the opposite side with the same magnitude of velocity. Think about the relative velocity of your car with respect to the truck and vice versa.

For the example that is described in the lesson, imagine that someone else is moving with the same the velocity as that of the wind; calculate the velocity of your car with respect to that someone else.

You are driving your car 40 miles per hour from south to north and the wind happens to blow from west to east at 30 miles per hour. Calculate the velocity of your car with respect to a person standing on the ground. Calculate the velocity of your car with respect to the wind. Think about the difference between the two cases.

Find the two most common coordinate systems in physics. Define the two coordinate systems and think about the reason or reasons that it is easier to understand the orthogonality in Cartesian coordinates.

Day 10
Motion in the vertical direction

You hold a coin and then release it. What
does the motion associated with the coin
look like?

You hold a coin and release it. What happens then? The increase in velocity is affected by the size of the gravitational acceleration. In essence, that is all that we need to know when studying a motion in the vertical direction.

You hold an object and realize that it accelerates because of the Earth. That is what a motion in the vertical direction is about.

Well, then, we will just go over a bit more detail since understanding the motion is going to be highly important for us to understand the projectile motion, something to be covered in the next lesson.

What you have studied in the two previous lessons is about understanding a motion in the horizontal direction with respect to the surface of the Earth; well, in a small scale. Here, we will go over the motion one more time but focus on the motion in a vertical direction instead. Well, this lesson is where a lot of students start getting lost when studying kinematics, so let us focus hard on this one before moving on.

What we have studied in the three previous lessons can be applied to all kinds of motion, in any direction. They are generic. We came up with equations that describe the final velocity and that of the final position for an object as a function of some quantities including the initial velocity, initial position, and time. In other words, whether we deal with a mechanical motion in a horizontal or a vertical direction, the generic form of the equations can be something that we can start with. So, you may wonder why we need to study the motion in the vertical direction separately from that in the horizontal direction. Answer: When we describe a motion in the horizontal direction, we generally do not worry about gravity. Approximately speaking, gravity does not have an impact on that in the horizontal direction. On top of that, in the horizontal direction, we assume that there is no other object but the object of our focus. In other words, we have no other entity that causes the object of our focus to get accelerated, but we just assume that the motion was taking place some- how. For that reason, we just want to cover a special case further and that is about understanding the motion in the vertical direction where the presence of the Earth is going to have some impact on the motion. It is nothing more than that.

How? Earth pulls objects down to the ground if it is at a distance. The presence of Earth causes the object to accelerate toward Earth. Furthermore, as far as the distance being much smaller than the size of Earth, we can assume, in most cases, that the rate of acceleration is almost the same. So, how much is the rate of acceleration? It is exactly the size of the gravitational acceleration. Objects in the air are going to gain velocity at a rate of the acceleration. For instance, you are holding a coin. The velocity associated with the coin at that moment is zero. However, as soon as you release the coin, it is going to "fall" down toward the surface because as it falls toward Earth, the rate of velocity increases by about 10 m per sec. In other words, the velocity is zero to begin, then 10 m per sec in 1 sec, then 20 m per sec in 2 sec, and so on. Again, the size increases by that much as long as the object stays in the air. The object is much smaller than Earth, and the distance is much smaller than the radius of Earth. Figure 10.1 illustrates the point well.

At this moment, you may wonder "why" it is so. Well, for now, we are not going to worry about why it happens but just accept the result. You will see the point clearly when you study the dynamics part.

You may wonder how we take the object gaining velocity into account. Let us start with the velocity equation. The equation is this:

$$\text{Change in velocity} = \text{Acceleration} \times \text{Time}$$

If you remember the one that you studied in the previous lessons, that is how the change of velocity can be written as a function of the rate of acceleration and time. Yes, we do need to assume that the rate of acceler- ation while in motion stays the same. So, for instance, someone is driv- ing a car and it accelerates at a rate of 100 miles per hour squared, that is going to be the rate of acceleration, and it will remain constant in the expression above.

What happens to the expression when dealing with a motion in the vertical direction, particularly one where we hold something to begin with and release it? We first assume that the rate of initial velocity is zero, so we take that out from the velocity equation. Then, as for the acceleration, we assume that Earth causes an object to accelerate at a certain size: 10 m per sec squared as an approximated value. Then:

Final velocity = Gravitational acceleration × Time

In other words, the size of the velocity now can be calculated as long as we know how long the motion takes place or vice versa.

Understanding the equation is going to be highly important when studying a projectile motion. You are going to see this point in the next lesson. In short, the rate of change of the velocity per time is going to be the size of the gravitational acceleration. For instance, after 4 seconds from the time when the coin is released from my hand, the velocity is going to be 40 m per sec. After 10 seconds, it is going to be 100 m per sec. In short, you simply take the gravitation acceleration and multiply the time. That is how we describe an object moving in the vertical direction that is not bothered by any other object but Earth.

On the same token, we go through the same procedure with the position equation. We are going to end up with:

Displacement = 0.5 × Gravitational acceleration × Time × Time

This is commonly known as an equation for the motion in the vertical direction. It happens to be very convenient for us to estimate the size of the displacement when something is held and released in the air. Do note that we are going to study why the size of the gravitational acceleration caused by Earth is such and such a size later in the dynamics part, so let us not worry about that part too much at this moment. We are going to study the "why" part later on. For now, let us just keep in mind that the gravitational acceleration is something that we need to consider when studying the motion in the vertical direction, whereas we do not for the motion in the horizontal direction.

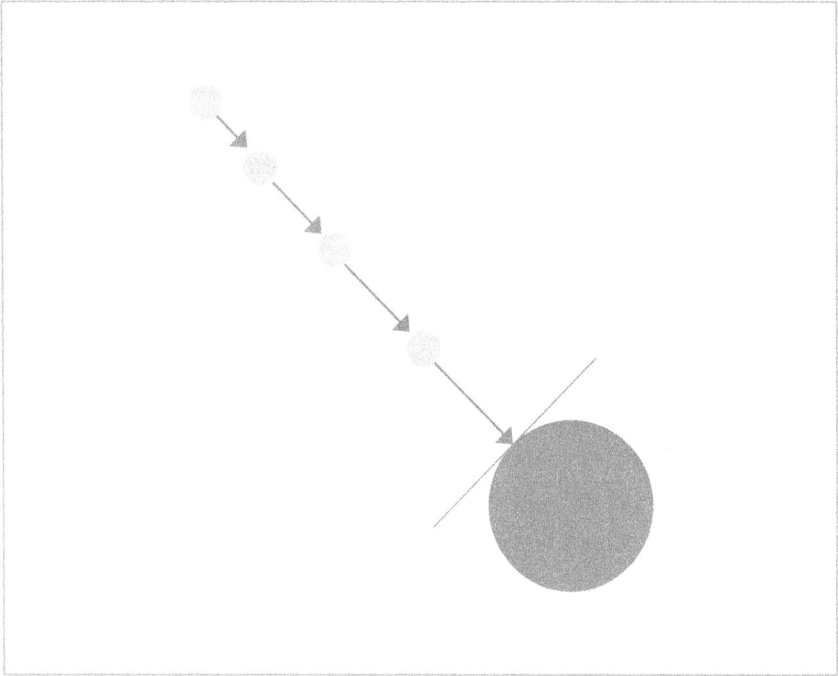

Figure 10.1: Imagine that we have an object and someone holds it at a distance with respect to the surface of the Earth. The object is light gray in color, and Earth is gray, and the thin line represents the line perpendicular to the radius of Earth. The object is released at a distance, and it is going to be pulled by the gravitational interaction with Earth. This is what we understand in most cases as a "freely falling" motion in the introduction to classical mechanics. Furthermore, the velocity, which is represented by the arrows in gray, grows with the gravitational acceleration.

Remember: what we studied in this lesson is somewhat related to the dynamics part. We are not dealing with a case where we have a single object in the system that we focus on, but we have Earth in the system as an additional entity instead, so we do deal with two objects in a system; thus, it is about dynamics and not entirely about kinematics. We did not study why the size of the acceleration is such and such, but we just take it as fact and focus on how the velocity and the position equations can be revised to accommodate the presence of Earth and how they are related to what we are going to study in the projectile motion.

Problems:

You hold a coin. You keep the distance between the coin and the floor at 100 cm and then release the coin. Calculate the velocity associated with the coin in the vertical direction just before it is going to hit the floor. Assume that we do not worry about any other objects being present but the coin and Earth.

You hold a coin. You keep the distance between the coin and the floor at 1 mile and then release it. Calculate the change of velocity associated with the coin in the vertical direction per second. How large is it, and why is that?

Day 11
Projectile motion

You throw an object in the air, and it is
going to fall to the ground. How would
you describe the motion in terms of all
the quantities that we have studied so far?

If you understand the basics of a projectile motion in classical mechanics, then you can proudly say loudly that you do have some understanding of what motion is about in classical mechanics. It is important to understand projectile motion. It deals with a motion associated with an object or objects in two- or higher dimensional space. The things that we'll cover may not be all that intuitive, so let us spend some time going over some of the details one by one.

Again, it all starts with something in motion, where the change in position is realized over time. Imagine that someone throws a ball in the air with a certain velocity and angle with respect to the horizontal direction. Question: can we find the size of the displacement of the object in the horizontal direction and how high the ball went up in the air from the size of the initial velocity? In other words, can we find how far the ball moves and how high the ball goes in the air? Understanding the relevant procedures to find the answers is the most important part that you need to get familiar with when studying a projectile motion, and that will certainly lead you to a better understanding of kinematics overall. After that, you can go over more complicated questions. Or the complicated questions will come to you naturally as you get more interested in studying physics.

At this point, please go over all the figures in this lesson first and come back here. It is going to help you understand the main point better.

Let us go over the procedure in more detail. When the velocity and the direction are given, what we can do first is calculate the velocity in both the horizontal and the vertical direction, which we can do with the initial velocity as a vector, something that we covered in the first two lessons. For instance, when the velocity is 100 m per sec and the direction with respect to the ground surface is 30 degrees, we can find the velocity component in the horizontal and vertical direction. Yes, we can. Have a look at Figures 11.1 and 11.2 if you have not done so.

For horizontal velocity, you need to multiply the velocity component by the cosine of the angle, and you do the same but that by the sine function for the vertical direction. Note: Make sure that you know how to get the horizontal and vertical direction component. This is where things are getting more confusing for those who did not understand the vectors well, so let us pay lots of attention here. You may want to read the vector lesson one more time. When studying undergraduate classical mechanics, it is easier to understand and approach different types of practice questions once you calculate the size of physical quantities in horizontal and vertical directions. In other words, you calculate vector quantities in the Cartesian coordinate system and that will make the rest of the work easier, as opposed to your playing with the physical quantities in general. For instance, for 100 m per sec with a 30-degree angle, the component in the horizontal direction is about 85 miles per sec and in the vertical direction is 50 m per sec.

What do we do next? What other information can we utilize? Yes, we just need to calculate the "time" during which the ball is going to be in the air. How? We need to utilize the velocity equation in the vertical direction. We know the initial velocity in the vertical direction, and we know the rate of change of velocity in the vertical direction caused by the gravitational pull, so we can calculate the time as follows:

We started with 50 m per sec and gravity pulls the object, so the rate of reduction in the velocity is about 10 m per sec. It takes about 5 seconds for the object to have zero velocity in the vertical direction. What is the rate of reduction in velocity? Answer: It is exactly what acceleration is in physics.

Figure 11.3 illustrates the point, and you will see the basics of what the "pulling down" is about in some of the lessons in dynamics part.

Then what do we do? Now we know that the object stays in the air for 5 seconds. We can calculate the displacement in each direction. For the total displacement in the horizontal direction, all we need to do is to multiply the time by the velocity in the horizontal direction, and for the vertical direction, utilize the position equation, where the displacement is written as a function of the initial velocity, time, and the acceleration, which is the size of the gravitational pull. How do we calculate the size of the displacement in the vertical direction? Have a look at Figures 11.4–11.6. All the procedures described so far in this lesson are represented by a few figures, so please go over them and ensure that you understand the procedures.

You will find many different types of problems that deal with the motion in two- or higher dimensional space, so we are not going to go over them all here, but the one that we covered is, again, one of the basic questions that you are encouraged to understand in order to solve other types of problems. Other types of problems are, practically speaking, variations of what we covered.

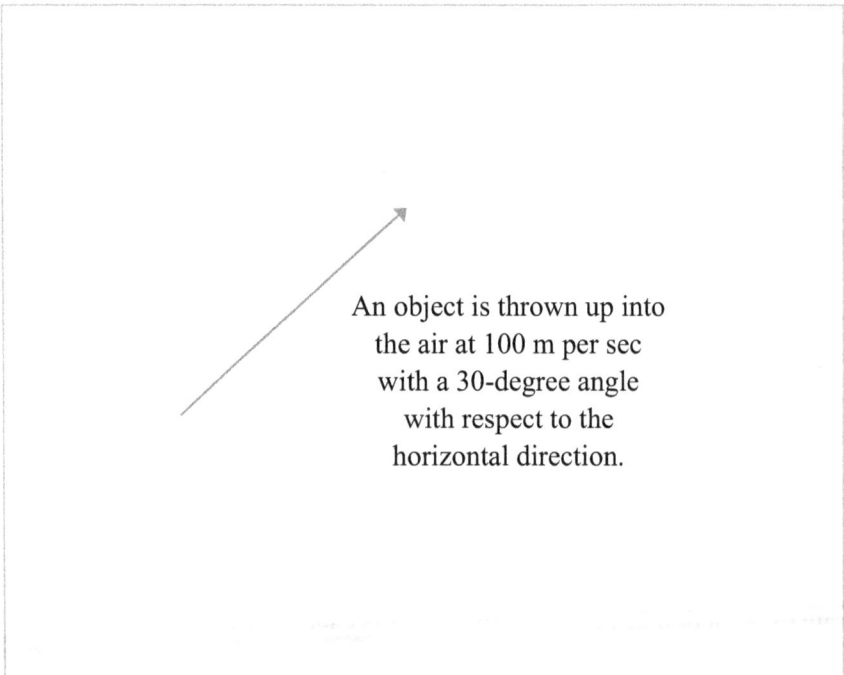

An object is thrown up into the air at 100 m per sec with a 30-degree angle with respect to the horizontal direction.

Figure 11.1: Let us further our understanding of a common type of projectile motion, a motion in two-dimensional space. Someone throws an object into the air with a velocity of 100 m per sec with a 30-degree angle with respect to the horizontal direction. Assume that the gravitational acceleration is about 10 m per sec squared.

100 m per sec
as the total

50 m per sec in
vertical

85 m per sec in
horizontal

The horizontal component
is about 85 m per sec and
the vertical component is
50 m per sec. Calculate the two
using trigonometric functions.

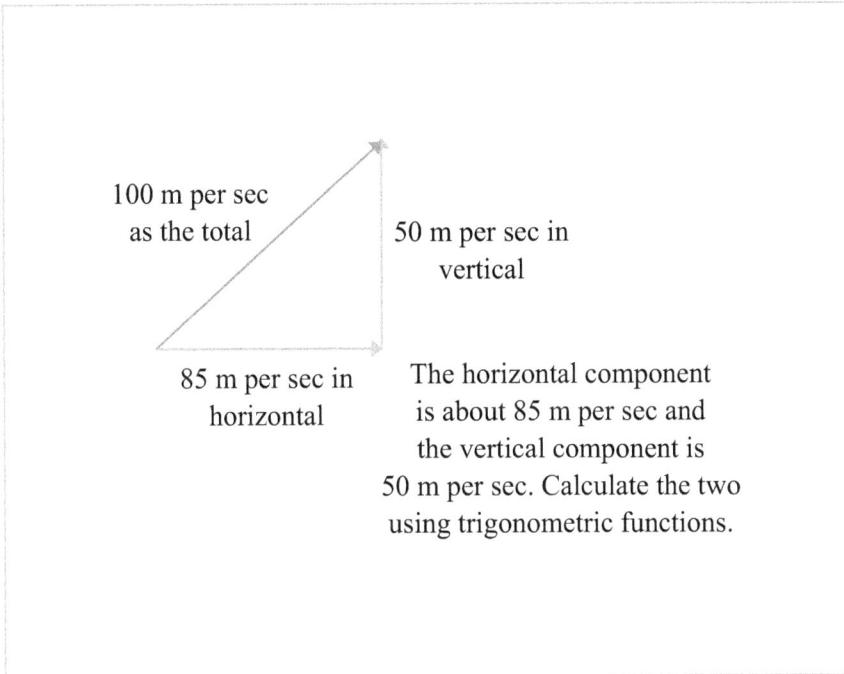

Figure 11.2: This is important. The velocity 100 m per sec with 30 degrees in the Cartesian coordinate system can be considered the same as the sum of 80 m per sec velocity in the horizontal direction and 50 m per sec in the vertical direction. You can get the size in each direction using trigonometric functions, and I am going to leave that up to you to study on your own.

10

20   Assuming that the
gravitational acceleration is
30   10 m per sec squared, the
velocity is going to be reduced
40   by 10 m per sec. Question: how
long does it take to get to the
point where the velocity in the
vertical direction is zero?

50

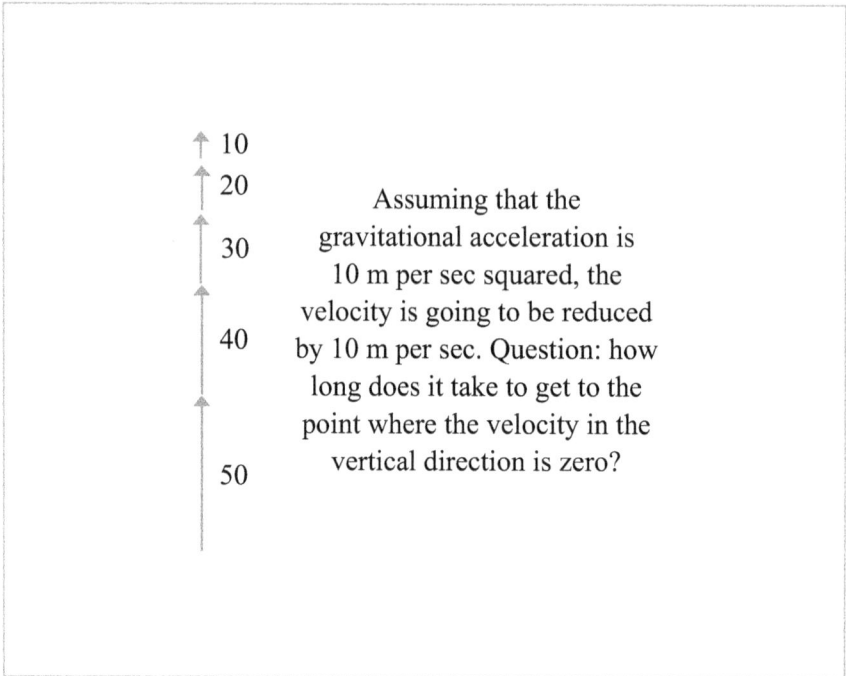

Figure 11.3: When we study the motion in the vertical direction, the
bottom line is this: all the changes happen in the vertical direction.
Assuming that the velocity is going to be reduced by 10 m per sec in
the vertical direction, how long does it take for the object to have zero
velocity in the vertical direction? How long does it take to get back to the
ground level from the time at which the object was thrown in the air?

The velocity in the
horizontal direction is 85 m
per sec, and the motion takes place
in 5 seconds. The gray arrow
in the figure shows the velocity
in each second for the first half.
How far is the object going to be
displaced in the horizontal direction?

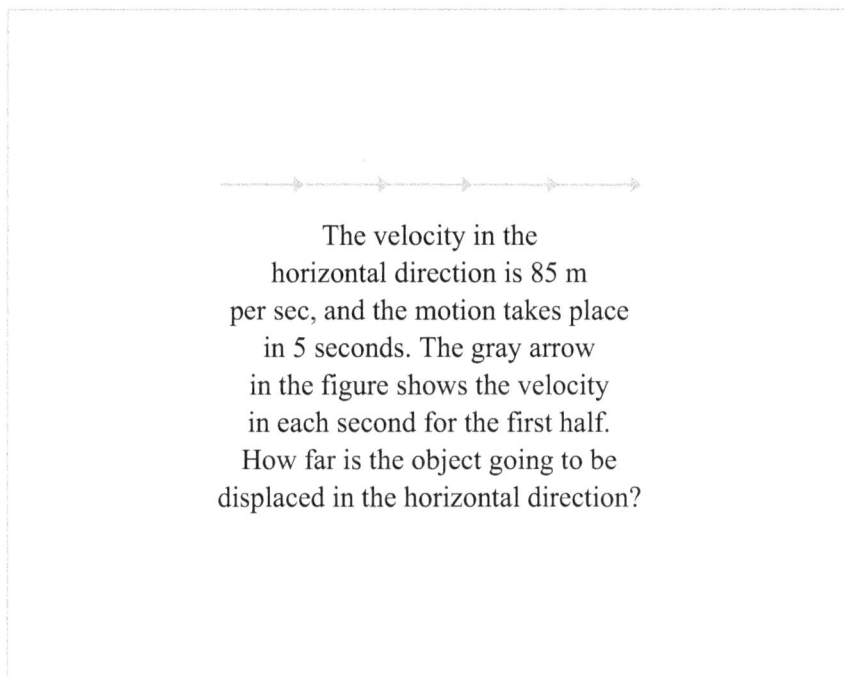

Figure 11.4: This one is important to understand, so I am going to
encourage you to take some time on this figure. What is changing in
the projectile motion is the velocity in the vertical direction. Again, the
velocity in the vertical direction changes. What does that tell us? That the
horizontal direction is going to stay the same from the beginning to the
end. Hint: there is no entity bothering the motion in the horizontal direction, and that is why. Each arrow represents the velocity in the horizontal
direction during the first half of the motion, and they all look the same.
Question: how far is the object going to move in the horizontal direction
in 10 seconds if the set of arrows in the figure represents half of the horizontal motion? Imagine that is the time during which the object is going
to be up in the air.

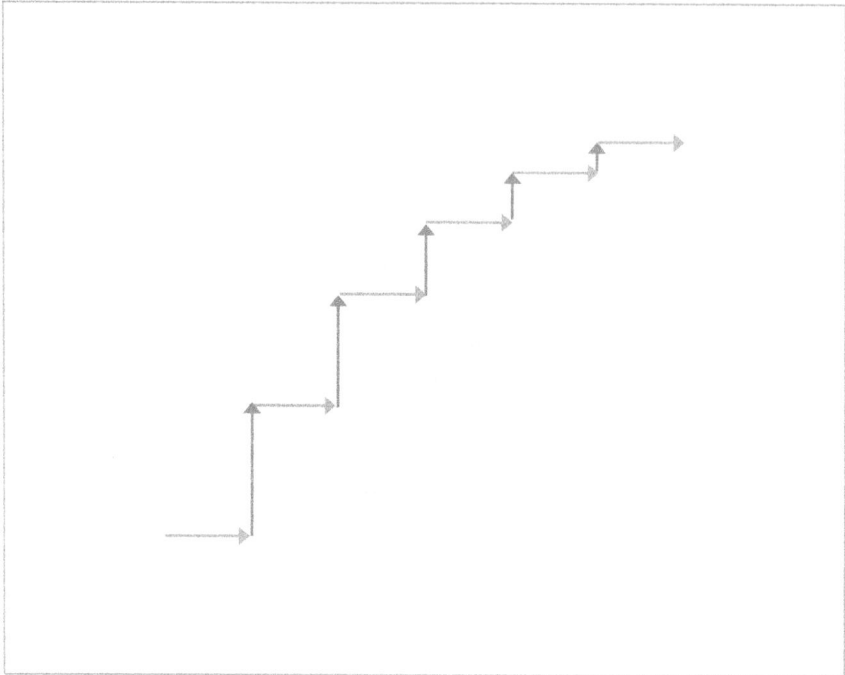

Figure 11.5: This is an interesting and important point. Let us combine the previous two figures, at least the first half of the motion; the light gray represents the velocity in the horizontal direction and the dark gray represents it in the vertical direction, where the size of the velocity reduces by 10 m per sec because of the gravitational pull of Earth. The figure shows the combination of the velocity in the two directions for the first half of the motion. Question: what does it look like? If you connect the midpoint of the light gray arrows in the figure, how does it look? Does it ring a bell?

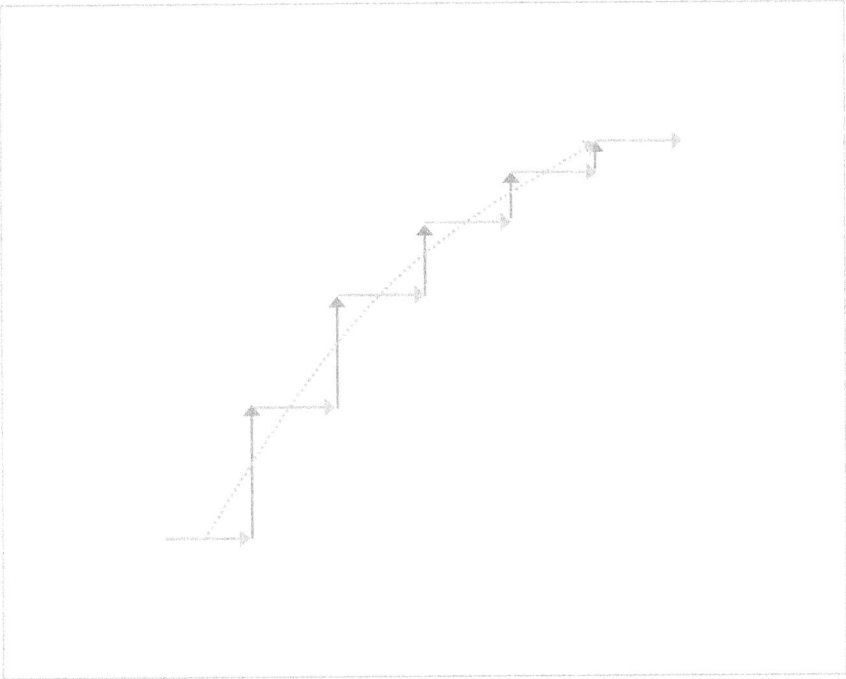

Figure 11.6: Following up on the previous figure, if you connect the midpoint of the arrows in light gray in the figure, you are going to end up with something like the grey parabola, which is the first half of the projectile motion. Now, how does that look for the second half of the projectile motion in this case? In other words, from 5 to 10 seconds in the example above, what does the motion look like in space? Furthermore, if you combine the first and the second half of the motion, what does that look like? I am going to leave that part up to you.

Remember: You calculate the velocity of an object in the vertical direction properly and then calculate the time during which the object stays in the air. If you do so first, then you are almost done with the projectile motion already.

Problems:

Please go over all the figures in this lesson one by one, and make sure that you understand the procedures.

Do the calculation for the example illustrated in the lesson one more time but with the size of the gravitational acceleration being reduced by a factor of two.

Someone throws an object up in the air with a certain velocity and angle with respect to the horizontal direction. Calculate the size of the angle at which the horizontal displacement of the object is going to be maximum. In other words, calculate the size of the angle at which the size of the displacement in the horizontal direction is going to be at its maximum when the size of the velocity remains constant. Assume that there is no air resistance or any other factors, but because of gravity, the velocity is in the order of 100 miles per hour.

What is covered here works when you calculate the size of the displacement for an object thrown into the air at 100 miles per hour, but it does not work well if the size happened to be about 100 million miles per hour. Think about why that is so and describe your reason or reasons. Hint: you may want to think about the size of the gravitational acceleration when you are the radius of Earth away from the surface.

Day 12
Unit conversion

Jae: Does anyone know the radius of a
quarter dollar?
Adam: Yes, I measured it, and it is
about 1.25.
Sarah: No, it is about 0.5.
Adam: No, it is 1.25.
Sarah: No, it is not.
Jae: Wait a second. Are we comparing an
apple to an orange? Are you talking about
the same thing in two different units?
Adam and Sarah: …

We set units for our convenience.

Imagine that you are taking a classical mechanics course. As an
assignment for your lab, you are asked to measure the size of a quarter
dollar. You are asked to measure the radius of the coin. Question: how do
you measure the size? How do you report the size?

Certainly, you could use a ruler to measure the size of the coin.
A ruler is readily available when you go to a convenience store. You just
buy it, and then place a ruler near the object and measure the size. It is
not that difficult. Or you may use a measurement app that you can down-
load on your phone. There are many different ways to measure the size of
the coin, and we have many methods of measurement readily available
these days. Point: we can measure the size. However, we need to be a bit
cautious when reporting it, even for a lab assignment.

We can measure the size, but we just need to be a bit cautious when
reporting it.

The issue is something to do with the unit: in which unit do you want to
report the size? Given the size of the coin, you may think of reporting it
in either cm or inch, the two common units. Sarah was one of students
in the lab; she bought a ruler that goes by the unit inch, placed it near
the coin, measured the size, and noticed it was about 0.5 inches. Then

she reported the size to the class saying that the radius of the coin was 0.5 inches.

So did Adam. Adam was another student in the lab, but he happened to buy a ruler that goes by centimeters, and he did the rest the same as what Sarah did. He measured the size to be about 1.25 cm.

The very next day, Sarah and Adam met in the physics lab, and they reported what they did. The conversation they had was illustrated at the very beginning of the lesson. Well, it is obvious that they measured the radius of the same coin, but they reported the size with two different numbers. What is going on?

Answer: They were reporting the same physical size in two different units. That was what was going on. Adam measured the size in "cm," whereas Sarah measured and reported the size in "inches." Furthermore, there was an issue when they were reporting their results. They did not specify in which unit they measured the size of the coin. They just assumed that everybody used the same unit. In the conversation, they were just saying the numbers but not specifying in which unit they measured the size. That was where things went in the wrong direction. They were comparing an apple to an orange; they were comparing their results in two different physical units. Believe it or not, the example above happens a lot in physics classrooms. Knowing the unit by which a quantity is reported is important. You will see the importance of the point again when you study the size of an angle in a rotational motion.

Coming back to measuring the size of the coin, well, how should the conversation be going, ideally speaking? It would be something like the following:

Jae: Does anyone know the radius of a quarter dollar?
Adam: Yes, I measured it, and it is about 1.25 cm.
Sarah: No, it is about 0.5 inch. I checked the unit on my ruler, and it was going inches.
Adam: Mine was in cm.
Jae: Okay, then, we are comparing two numbers in two different units. What do we need to do then to compare the two numbers?

> Sarah: Well, we need to be more
> consistent.
> Jae: Yes, but how?
> Adam: We convert the unit.
> Jae: Yes, that is correct. 1 inch is about
> 2.5 cm. So, Sarah, your 0.5 inch can
> be expressed as 0.5 of 2.5 cm. In other
> words, you multiply 0.5 by 2.5, then you
> end up with 1.25 as your measurement,
> and the unit is in cm.
> Sarah and Adam: Oh, I see.

So, again, the point is that Sarah and Adam were comparing two different numbers in two different units. When the number in inches is converted to centimeters for Sarah, she can report that her number is the same as what Adam reported as 1.25 cm since 1 inch is about 2.5 cm; the conversion factor is about 2.5. Therefore, when she multiplies 0.5 by 2.5, she is going to end up with 1.25 as her number, and the unit associated with the number is going to be centimeters. Again, when they compare the two numbers, they need to compare them once they express their numbers in the same unit. We need to compare an apple to an apple, not to an orange, especially when it comes down to the unit.

Here comes an interesting question: does that mean that they need to measure the numbers in the same unit all the time? Answer: No, they do not. They do not have to. As you see in the conversation, they can measure the two numbers in two different units. It is just that when it comes down to comparing the numbers, they need to convert the results in a unit to another unit so that the two numbers can be compared in the same unit. Point: as long as you are using the same unit and specify the unit in your report, you are okay with reporting the number. It does not have to be a specific unit. You can always do a unit conversion later. It is just a bit more convenient to play with a single unit from beginning to end and not worry too much about which unit needs to be used. In other words, you just need to be consistent.

You just need to be consistent with using a physics unit. It does not matter which unit the quantity is going to be reported in as long as the dimension associated with the unit matches.

Here is another interesting question, and many students get confused at this point: is there another unit by which we can report the size of the coin? In other words, can I come up with my own unit? Answer: Yes, we can report the size in any unit that we want. You can certainly set your own unit as you want. Some students think that inches and centimeters are all we have when reporting the length. However, it does not have to be centimeters or inches. They just happen to be the most widely used units when measuring the size of a coin or something with a similar dimension. Again, we can set our own unit but with one condition.

For instance, you measure the size of the radius of the coin in the unit of the size of the nail of your "thumb." In other words, you are going to use your nail as a unit. You place the coin next to your thumb, and you roughly measure the size of the coin to be about 1.5 times of your nail. Given that, you can report the size of the coin as 1.5 "nail," the name for a unit that you come up with on your own. It is okay to measure and report the size that way but, again, with an important condition: the size of the "nail" does not change over time, and no matter how many times you measure the size of your "nail," you can be consistent with it and people will widely accept the unit by using some sort of medium that is readily available. The size of the nail should not change over time, so people can use the size as a reference. Now, you may sense what we are heading toward.

Physical units just need to be consistent and widely acceptable.

Then we can use those units. As long as the unit can be readily replicated and available, we can go with any unit that we want. The same goes for units in time and mass. We just need to be consistent when using the units when working on your problem.

Remember: We set a unit for our convenience. We can set our own unit and use it as long as the size associated with the unit can be readily and consistently replicated. For instance, if you want to measure the size of your gravitational acceleration in feet per sec squared instead, it is certainly okay to do so. You just need to use feet as a unit through-out the whole problem that you work on. You can convert the unit later if need be.

Problems:

Calculate the size of the gravitational acceleration on a small object that is due to the Earth in feet per second. Can you do it or not? If you can, why? If not, why not?

Adam measured the radius of the coin to be 1.1 cm and Sarah measured the same to be 0.5 inch. After studying the lesson, Adam converted his number to inches and they compared the two numbers. If the two num-bers are not the same, think of all the possibilities for the two numbers being different.

CHAPTER 2

Dynamics

We study what happens when we have two or more objects within a system. Whenever we have more than one object with mass in a system, they are going to interact gravitationally to each other. We study their interaction. We study the cause for their being in motion.

Day 13
Cause and effect

> "When something is in motion, there
> must be a reason for it being in motion."

All the previous lessons are important for you to understand classical mechanics better. That is why they were included in the book, but this one is something really special. Why? Because this lesson is going to give you an opportunity to get some sense of why we study physics and how different lessons are related to each other and how they are related to what are to be covered in other subjects, including electrodynamics and quantum mechanics and potentially other topics. So, let us focus closely on this lesson if that is okay with you.

So far, all we have studied is a motion associated with a single object. In kinematics, we assume that the object is the only object in a system and there is no other object or objects "causing" the object that we focused on to be in motion. In other words, we assume that the object has somehow already gained its motion. We do not think about or care about "why" the motion was initiated. In other words, we studied the kinematical quantities associated with the object. We did focus much on "how" and "where" the kinematics were coming from.

However, the situation is a bit more complicated than that in reality. Why? It may already be obvious to some of you if you read the previous lessons with focus. Think about the number of objects. We have more than one object in this world, and no matter how hard you are working, it is difficult to realize a system where you have only one object being present. You are most likely to end up with two or more objects in a system.

What is my point? The point is that the objects in a system are going to "interact" with respect to each other; they are going to have some impact on other objects in the system, and having such interactions is what "causes" an object to gain its mechanical motion. This is the essence of what cause and effect in physics is about. Summarizing it:

> We need another object that is causing
> the motion of another object. Kinemat-
> ical motion of an object or objects does

not come out of nowhere. The "cause" is
the presence of the other object, and the
"effect" is the motion associated with the
object that we focus on.

Let us go back to the topic and have a look at it from a different point of view. Let us assume that there is an object, the object is the only object in this world, and the object is in motion. As you can imagine, this is not a realistic situation, but, in any case, let us accept it as truth. Question: what is going to happen? Answer: The object is going to keep the same motion.

In other words, the motion keeps going on forever. It might be a bit hard for some of you to imagine a case like that, but think about learning how to ride a bicycle for the first time. Someone helps you at the start, but once you gain your motion, or your momentum, you keep the same motion, even if someone is not holding you from the back. When there is nothing interfering with you, the motion that you have already gained won't change. This can be applied to other mechanical cases too.

Why is that possible? Answer: Again, there are no other objects to "interfere" with the motion associated with the single object since it is the only object present. The motion is going to keep going forever. What we study in kinematics is somewhat relevant to the case: an object somehow gains its motion and we analyze it. That way, we can focus on the motion instead of worrying too much about the interaction between objects. We do not worry about the cause. However, note that we know that this is not what we deal with in our daily lives; there is almost always another object present in a system. We have more than one object in a system.

Question: what happens then? Yes, they are going to gravitationally interact with each other. When we have two objects in our system, one object pulls another object and vice versa. I am going to go over more details about the properties associated with the interaction in another lesson, so here let us just understand that there is going to be an interaction mechanism that pulls one object to another, as long as there is more than one object present in a space. Remember: when more than one object is in a system, they are going to interact with each other.

If you are really thinking, you may ask the following: What is the implication of having such an interaction mechanism? Believe it or not, we have already studied that a bit when we studied kinematics in this book. Assume that there are two objects in a system and that they are not moving but stay where they are at the beginning. For example, there is a pen in New York, and that is the only object in this world. The pen is going to stay in New York forever, from the beginning to the end. Why? There are no other objects affecting the motion of the pen. The pen is not bothered by any other entities. But the situation will be slightly different if there is another object present in the system. Let us assume that there is a piece of paper in Chicago, so now we have two objects in this world, a pen in New York and a piece of paper in Chicago. Given that the gravitational interaction is going to pull them toward each other, the pen is going to start moving toward Columbia and the paper is going to start moving toward Charlotte. They are going to start moving because of the presence of the other object. In other words, the pen gained its kinematical motion because there is a piece of paper and vice versa. Figures 13.1–13.3 clarify the point further.

As mentioned earlier, this is somewhat different from what we've studied so far in kinematics. In previous lessons, we analyzed the motion associated with the pen or the paper. But here we try to understand why such motions even start. We are studying why the velocity of an object changes and how it ended up with such a velocity and acceleration at a certain time. Then we need to introduce the presence of another object and force associated with the two objects. The pen moves and gains a certain velocity and acceleration after a certain time because of the presence of the paper, so there is going to be a gravitational force between them, and because of that force, the pen begins to move toward the paper; the paper causes the pen to move. In terms of the pen, the paper is the cause, and the motion of the pen is the effect. This will be what we study here in the dynamics part of classical mechanics. There is a "cause" for a motion.

Circle: I am a circle. It is nice to meet you.

Jae: But wait a minute. You are moving. But how? Is someone or something pushing you? I think that there should be a "cause" for you being in motion. Am I right?

Figure 13.1: This illustrates the point of the cause and effect. If "something" is in motion in a system, there should be something else that caused that motion. Saying it differently, the motion is not coming from out of nowhere, although we did not care about the cause part when we studied motion itself back in kinematics. Point: there must be a reason for "something" to be in motion. It is essential.

I have a pen in
Charlotte,
North Carolina.

You have a piece of
paper in Columbia,
South Carolina.

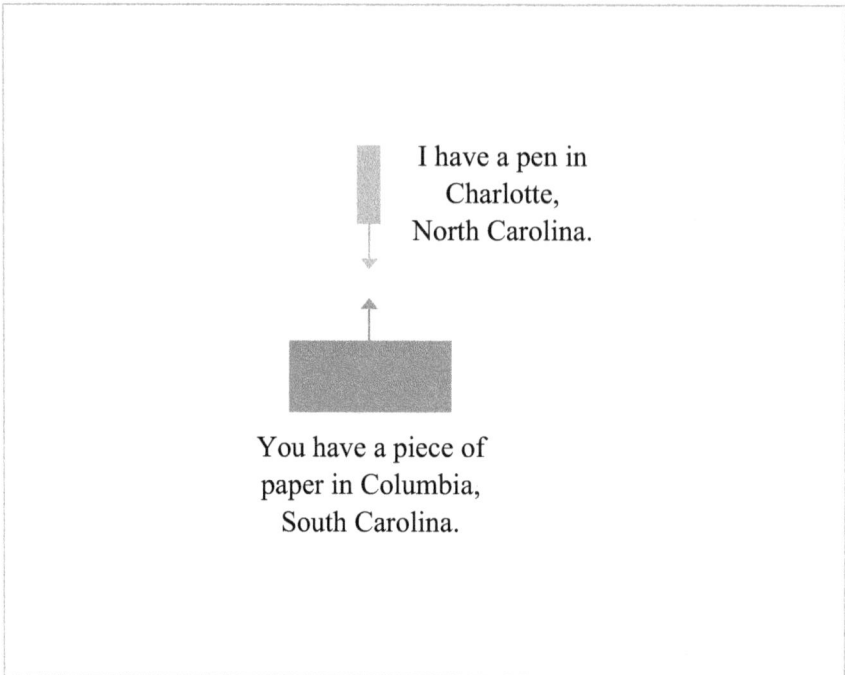

Figure 13.2: This illustrates the essence of two objects interacting with each other in classical mechanics. There is a pen in Charlotte and a piece of paper in Columbia. In other words, there are two objects in a system, and they are at a distance from each other. Furthermore, we assume that they are the only two objects in this system. Any two objects with mass are going to "pull" each other via gravitational interaction. It is an attractive force; therefore, an object will be pulled toward the other object, and the size of the pulling is quantified by gravitational force, following the universal law of gravitation. The force is proportional to the size of the mass of the objects and inversely proportional to the distance squared between them. Again, the direction of the force is always toward where the other object is, which is indicated by arrows in the figure. A pen is causing the paper to be in motion and vice versa.

There is a pen in Charlotte, and the pen is going to move toward a paper in Columbia.

There is a paper in Columbia, and the paper is going to move toward a pen in Charlotte.

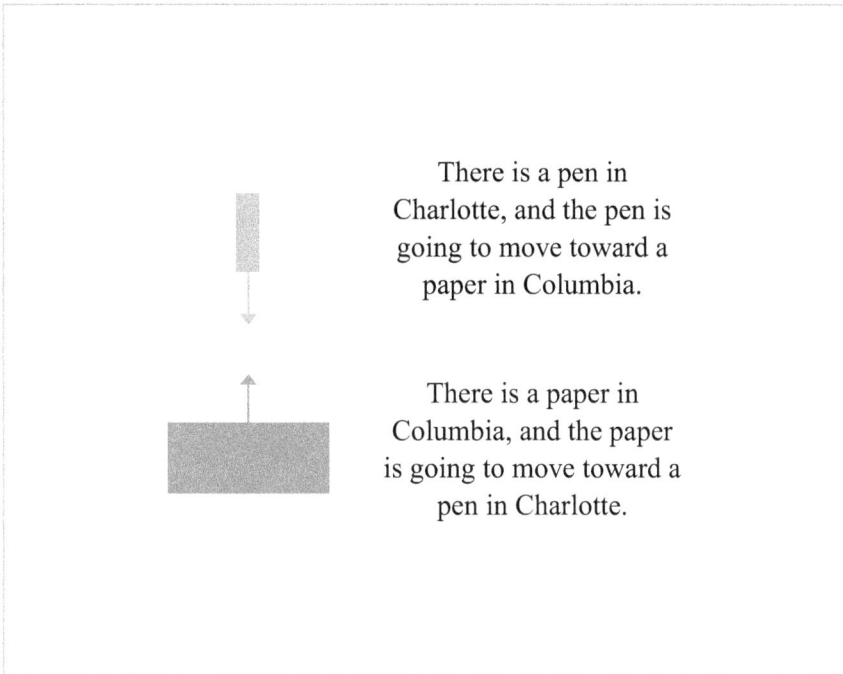

Figure 13.3: This is to add one more layer to the lesson. The figure is the same in essence as the previous one, but it covers something slightly different. If there is an action, then there must be a reaction. It is as simple as that. If the pen accelerates toward where the paper is because of the gravitational force between them, then the paper accelerates toward where the pen is because of the same force. "Force comes in a pair." There is no single force in physics, and that is so since the gravitational force is defined as a function of the mass of the two objects and the distance between them. That means that there is no such thing as an object being in a motion and gaining acceleration out of nowhere. There must be a reason. There must be a cause for the motion; thus, an object is going to gain kinematical motion. The principle of action and reaction is what is behind the reason.

Remember: this part of the lesson may not seem that important if you just want to ace your test, but this lesson is going to be highly critical to understand what really is going on in classical mechanics. If you don't understand how the kinematics and the dynamics parts are related to each other, you are going to have a hard time understanding what classical mechanics is about, so I am going to encourage you to come back to this lesson once you study all the lessons in the dynamics part.

Problems:

When we have a pen in the city of Charlotte and a piece of paper in the city of Columbia, they are going to gravitationally pull each other. Calculate the size and the direction of the acceleration of the pen in terms of that of the paper. The pen is 0.1 kg, and the paper is 0.01 kg. You may Google the distance between the two cities.

Imagine that you are going to create a system. Can you realize a physical state where a single object is present, and the object is going to be perfectly at rest with respect to the system as a reference frame? In other words, can you create a system where an object could be truly not in motion with respect to your inertial frame? If so, describe your reason or reasons. If not, why not?

Imagine that something is in motion with respect to some inertial frame, but none of the objects in the system are interacting gravitationally with respect to each other. Does that mean that all the objects in the system are going to be at rest? If so, describe your reason or reasons. If not, why not?

Day 14
Gravity

"Any two objects with mass at a distance
will always pull each other because of
their gravitational interaction. It is an
attractive force."

As mentioned in the previous lesson, any two objects at a distance are
going to pull each other because of their gravitational interaction with
respect to each other; it is a mutual process. Going back to the situation
where we have a pen in Charlotte and a piece of paper in Columbia, the
pen is going to start moving toward Columbia, and the paper is going to
start moving toward Charlotte because they are going to gravitationally
pull each other. That means that they are going to pull toward each other,
so the distance between them is going to get smaller as time goes on. So,
the bottom line is that a motion takes place because of the presence of
other objects in a system; an object being present causes other objects
to be in motion, or to be in a state that is different from the past. In other
words, if there is an object and it is not in motion to begin with, the
object is not going to be in motion, unless another object is present in the
system. It is the "cause" of the first object being in a motion.

Now let us shift gears a bit. If you think about it, what we have studied so far is about the direction associated with the interaction, strictly
speaking. We learned that they are pulling toward each other, so there is
information about the force. Do we know how strong or how weak the
interaction among objects is going to be? To answer the question in a
clear manner, it is time to introduce the universal law of gravitation, one
of the most important laws in general physics:

The magnitude of gravitational force is
proportional to the mass of the objects
and inversely proportional to the distance
squared between the objects. It acts on
both objects with a same magnitude but
in the opposite direction.

Figures 14.1 and 14.2 may help you understand the statement better. The size of the arrows in the figures represents the magnitude, or the size of the gravitational force. Figure 14.1 shows that the amount of force gets smaller when the mass of the objects gets smaller. This is so since the force happens to be proportional to the size of the mass following the universal law of in gravitation. Figure 14.2 shows the size gets smaller and smaller as the distance gets larger for the same objects. In other words, the smaller the mass, the smaller the force. The larger the distance, the smaller the force. The mass and the distance are the two crucial quantities from which the size of the gravitational force can be calculated.

Some examples might help you to understand the point better. For instance, one object has a mass of 1 kg, another has mass of 50 kg, and the distance between them is 5 m. You can find the mathematical expression for the size of the force in many sources in the literature, so let us not go over the details. If you do some calculations, the size of the force between them is 2 Newtons. We assume that the coupling strength of the force, the universal gravitational constant, is 1. On the same line of thought, the force between them is going to be 4 Newtons if the mass of one of the objects doubles. Point: the heavier the object is, the larger the magnitude of the gravitational force. On the same token, the farther the distance between the object, the smaller the magnitude of the gravitational force between them. Remember that objects are going to gravitationally interact with each other when they have mass. The degree of the interaction is proportional to the size of their masses and inversely proportional to the distance squared between them, and this is referred to as universal gravitational law.

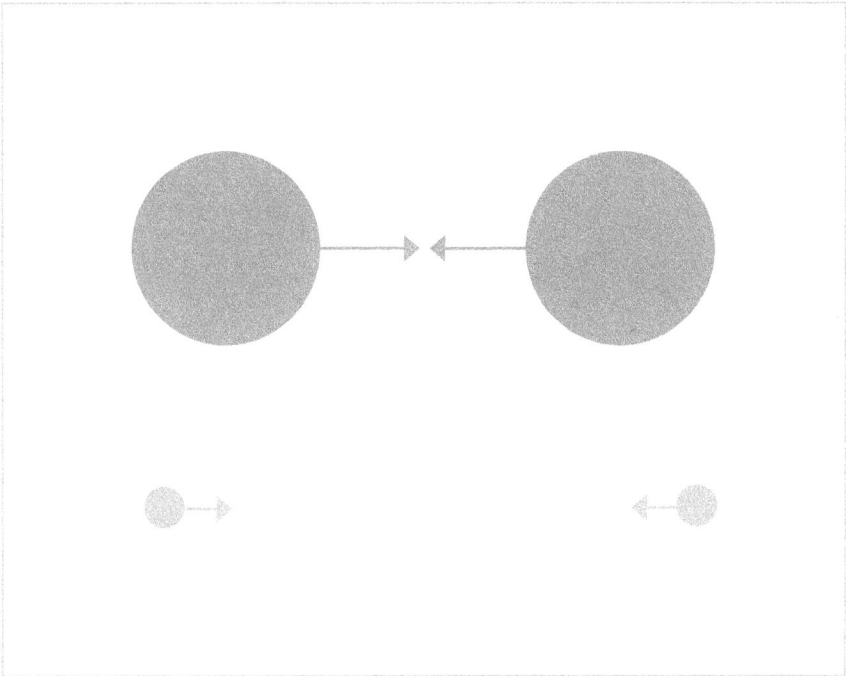

Figure 14.1: This illustrates the size of the gravitational force as a function of the size of the object. The smaller the mass of the object, the smaller the size of the gravitational force. The size of the arrow indicates the magnitude of the force. The objects on the bottom are smaller than the objects on the top so, assuming that they have the same density, the size of the force between the objects on the bottom is smaller.

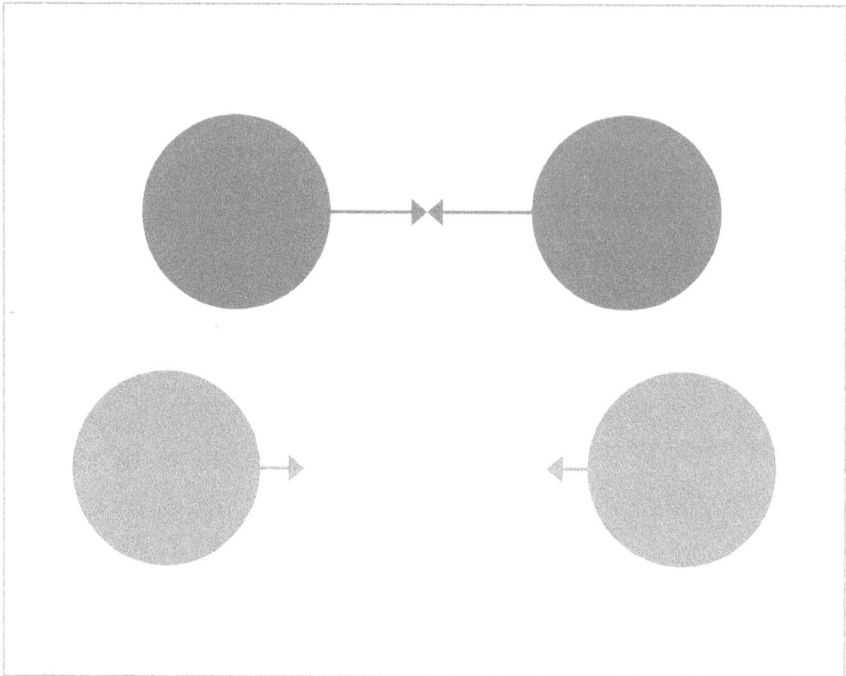

Figure 14.2: This illustrates that the size of the gravitational interaction depends on the distance between the objects. The farther the distance between the two objects, the smaller the size of the gravitational force. Again, the size of the arrows indicates the size of the gravitational force between the two objects, so there is a reason that the arrows on the bottom are drawn smaller than the arrows on the top.

Remember: gravitational force is only one of four types of fundamental forces that we have identified and studied so far. So, when objects are in physical motion, it could certainly be due to gravitational force but also to other forces. It does not have to be solely due to gravitational force. For instance, fundamental interactions such as electromagnetic interaction can cause an object to be in motion because of their interacting with each other by Coulomb force.

Problems:

"Gravity has nothing to do with Newton's laws." If it does, why? If not, why not?

Think of a situation where all objects in a system are not accelerating but are moving with a velocity that is the same from the beginning to the end. Describe a reason or reasons that such a state cannot last.

We know that we are on the surface of the Earth instead of being pulled down to the very center of the Earth by the gravitational interaction. Think about why this happens and write a paragraph on it.

Derive the size of the gravitational acceleration of a person near the surface of Earth. Briefly describe why it matters whether it is a person or a large object that is comparable to the mass of the Earth. You might want to work on this question after going over Newton's second law.

Day 15
It is mass

"It is mass through which kinematics
is related to dynamics, and it is also an
invariant quantity in classical mechanics.
It is that important."

Question: you have two objects in a system, and they are about to hit each other. How do you measure the impact on one object caused by the other one?

If you ask what mass is in physics, you may end up receiving different answers depending on the person or the people who answer the question. The answer may depend on how well a person "answers" the question too. The same answer seems to be described in a different manner. The bottom line is that mass is a very intriguing quantity in physics, so it might be hard to give you a clear answer to the question in a few sentences. For that reason, and because we do not intend to introduce many mathematical details, we are not going to cover the physical aspects associated with the mass in detail, such as what particle physicists do, but we are going to briefly describe its aspects to the kinematical and dynamical quantities that we study in undergraduate classical mechanics instead. That means we are going to address the properties and characteristics associated with the mass that are relevant to the materials that we covered so far. It is at least going to give you a better idea of what mass is practically about. For that, we are going to have a taste of three of Newton's laws.

Let us begin with Newton's first law. As we learned before, when no net force is acting on an object, the velocity associated with the object is going to stay the same forever, in a particular reference frame. Question: imagine that there are two such objects with the same shape but made of different materials so that the density of the two is going to be different. In other words, you can create a system where you placed the first object alone, take it away, and then place the second one in the same location. I know it is a bit unrealistic, but let us imagine that somehow this happens. If someone watches the two objects moving with the same velocity in the air, and no other objects interfere with them, can he or she distinguish the two objects? Answer: No, probably not. They have the same shape, and they're moving with the same velocity. In terms of their visual, we cannot identify the two differently. We simply do not have a practical method or tool to distinguish the two. This is an important point to get some sense: there is no need to distinguish them. We do not need to. Think about the statement for a moment, and then think one more time if it is not coming to you clearly. They just keep moving up in the air without interacting with another object. They are alone in the space. Can you think of a reason or reasons to distinguish the two? Well, we cannot. On the same line of thought, imagine that there is only a single object present in our universe. Do we need to know how heavy or light the object is? Does it matter or not? More importantly, can we even do that? As the title of the very first lesson in the book goes, everything is relative, but the term "relative" gains its meaning when there is something to compare. See how the lessons are correlated to each other?

Answer: So, it matters when there is another object to be compared with, but it does not when there is only one object. Figure 15.1 might help you understand the point better.

We do not need mass if there is only
a single object that we are studying or
analyzing.

Then, seeing the issue from another point of view, when do we need to distinguish them? It is only when they "interact" with other objects in a system, which is typically what happens when trying to analyze a mechanical motion. For instance, if one of the two is made of plastic and the other is made of metal, the latter is going to have more impact on the object that is going to be hit. Why? Yes, that is correct. Because it is "heavier." The mass is heavier than that of the former, so the latter is going to have more impact when interacting with others in a system. The heaviness that we are paying attention to, or the mass, now gains its importance if and only if it has something else to interact with in a system.

The heaviness that we are paying attention to, or the mass, now gains its importance if and only if when it has something else to interact with in a system.

That is one of the main reasons for the quantity "mass" being introduced in the dynamics part. It helps us to get some sense of how different the size of the "impact" is going to be when two objects interact with each other. Figure 15.2 illustrates the point here.

Wait a minute. Do you know what you just read in the above paragraph? Does that ring a bell? What is it really about? Yes, it is just what the principle of Newton's second law is about. Given the acceleration, how much impact can an object make on another one? That is what that is about. We are going to study that later though, but you just tasted it a bit, and we will go over the law further in another lesson.

Now, let us get to the last piece in Newton's laws. Yes, that is the principle of the action and the reaction. The first object is made of plastic, and it happens to be moving in a direction and it is going to hit, or collide in physics terms, with another object that happens to be at rest. Here is an important question. Do you think that there is going to be any difference depending on the type of material the object at rest is made of? I mean, depending on whether it is made of plastic or metal, will it make any difference in terms of the kinematics of the object after it collides with the moving object?

Yes, it does. There is going to be some difference. The one that is made of metal is heavier, thus the velocity associated with the object after the collision is going to move with a slower velocity. Why is that? Because the size of the reaction to the object at rest should be the same. Question: following which law? Yes, it is Newton's third law. The heavier the object, the smaller the magnitude of the acceleration is going to be given the same magnitude of force being provided to the object at rest. The principle of action and reaction tells us that the action needs to be the same as the reaction in terms of the size of the force.

Figures 15.3 and 15.4 illustrate the point here. The amount of force is going to be smaller for the interaction between objects in light gray. So, the impact on the object on the right-hand side is going to be smaller for the bottom one as shown in Figure 15.4. What is my point here? The point is that we do need mass to be introduced to properly illustrate the size of the gravitational interaction following the principle of action and reaction in Newton's laws.

> Neither a single action nor a single reaction can take place. They must be realized as a pair, always.

In summary, we just covered the physical aspect of mass in mechanics in terms of Newton's three laws. We are going to go over them again, one by one, but here I tried to explain how they are related to "mass" in physics. There are different aspects associated with mass, but I hope the lesson here gives you an opportunity to gain some knowledge of what mass is about when studying classical mechanics. If you are interested, I encourage you to study further what mass is about from the particle physics point of view. Where is the mass coming from?

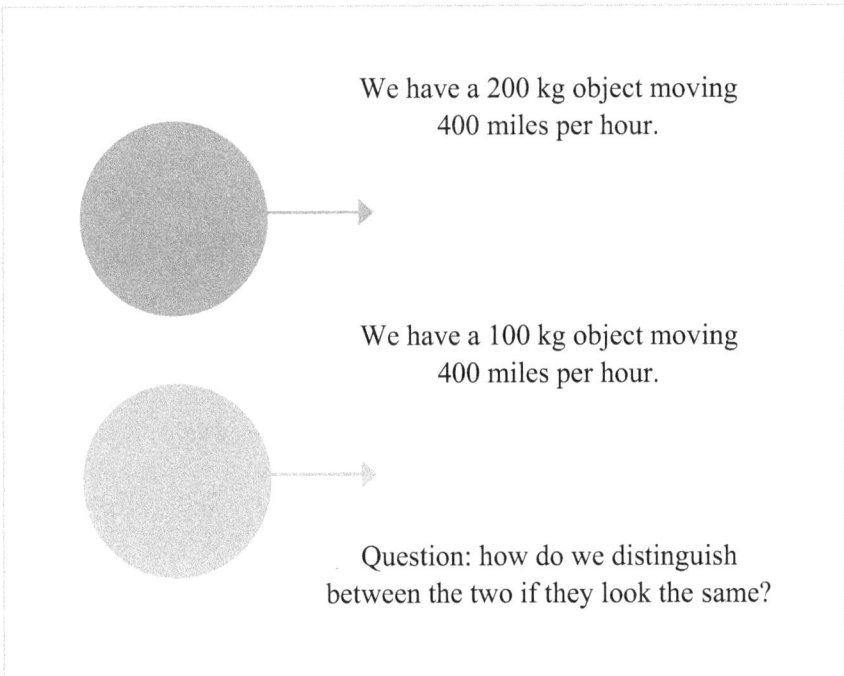

We have a 200 kg object moving 400 miles per hour.

We have a 100 kg object moving 400 miles per hour.

Question: how do we distinguish between the two if they look the same?

Figure 15.1: This shows that there is a need to introduce mass as a quantity in dynamics, when we have more than one object in the system. When we have two objects moving with the same velocity and with the same shape, and if the velocity is the only quantity that we may deal with, we may have an issue. We cannot distinguish the two objects in the figure. We need to introduce mass as another quantity in dynamics when we have more than one object to analyze. In dynamics, we study interaction among objects, so if two objects happen to have the same velocity, we need to distinguish their motion by introducing mass as an additional quantity to be analyzed. Otherwise, we cannot distinguish between the two.

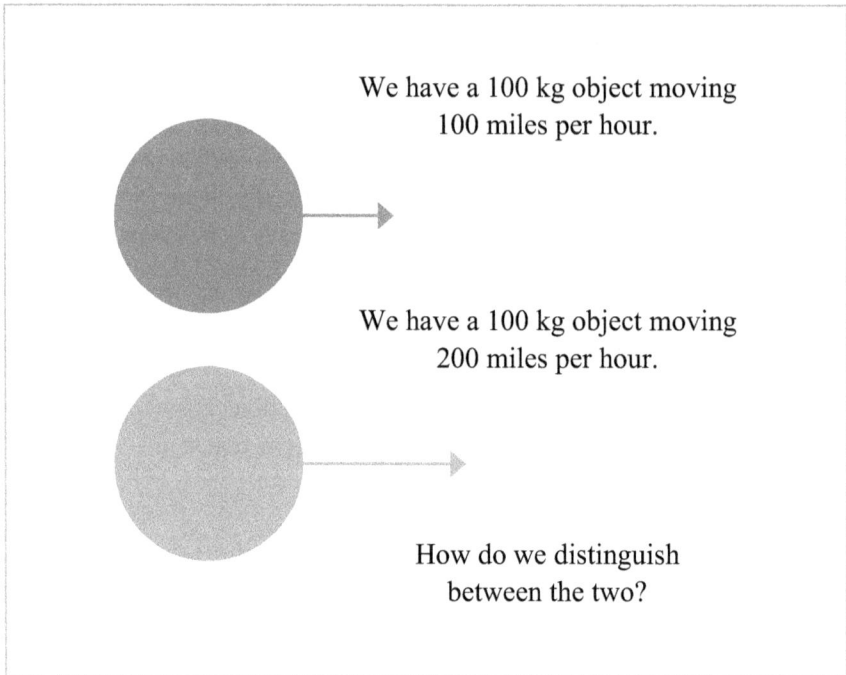

We have a 100 kg object moving
100 miles per hour.

We have a 100 kg object moving
200 miles per hour.

How do we distinguish
between the two?

Figure 15.2: This looks the same as what was shown in the previous figure but with the same mass and a different degree of velocity. So, this is an opposite case from the last one. Question: how do we distinguish between the two? Can you think of what dynamical quantity needs to be introduced?

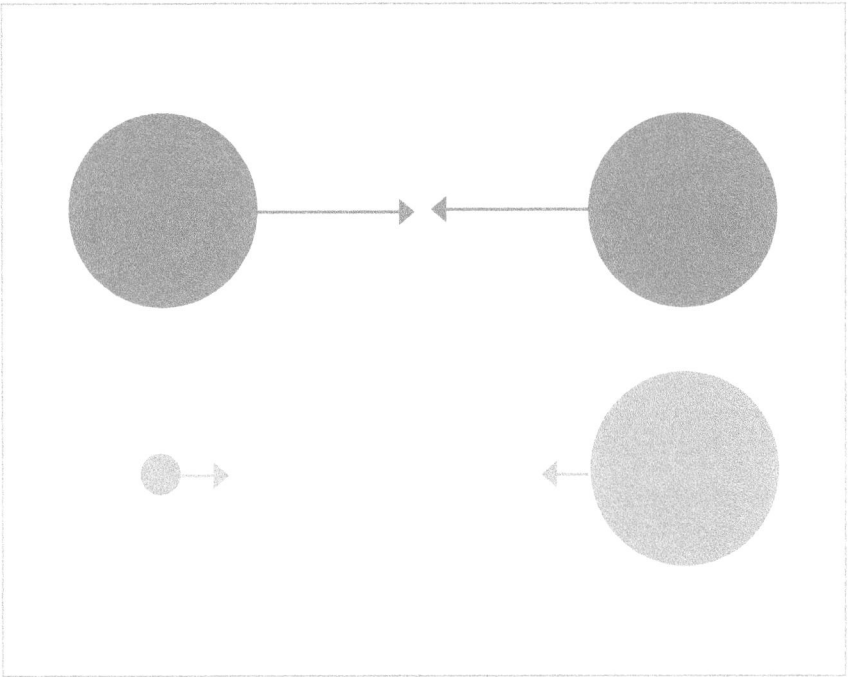

Figure 15.3: This illustrates that the size of the gravitational force is smaller between the objects in light gray since the one on the bottom left is smaller than the one above. We assume that they all have the same density.

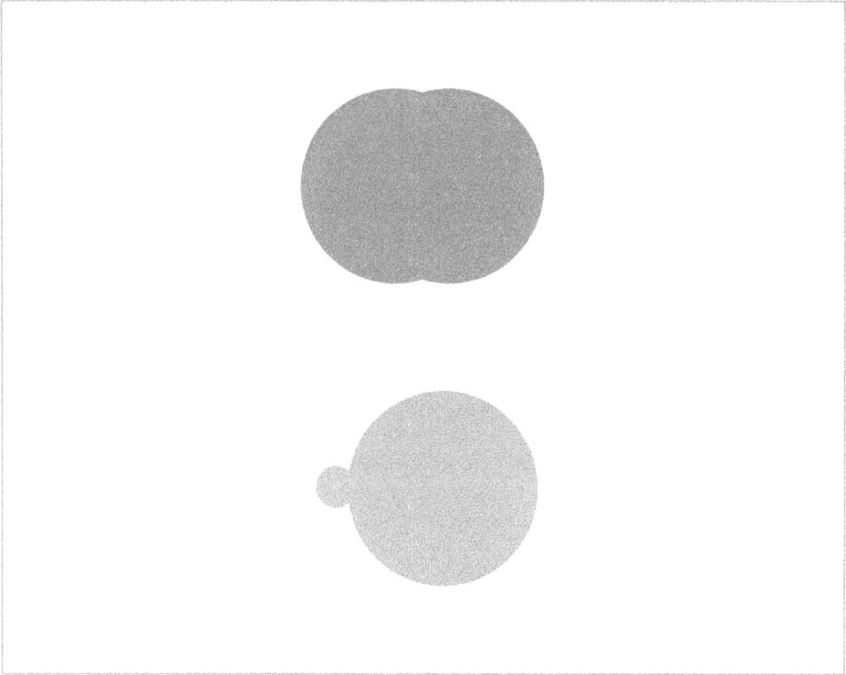

Figure 15.4: Since the object on the bottom left is smaller, the impact by the object in light gray is going to be smaller, following the principle of action and reaction. It is another reason for us to introduce mass when describing dynamics in classical mechanics. Without mass being introduced, we cannot properly describe the action and reaction. We cannot distinguish the interaction on the top from the bottom, assuming the two objects have no dimensions.

Remember: in classical mechanics, mass is a critical quantity to quantify mechanical motion, but it is also one that cannot be directly measured if you think of it. We can measure acceleration by physically measuring change in distance though.

Problems:

Think about why we did not introduce mass as an important physical quantity when studying the kinematics part, where we analyze the mechanical motion associated with a single object but in the dynamics part. Hint: do we need to know the size of the mass when there is a single object in our universe?

Read some books or articles that explain the test that Galileo did, and the principle associated with the test, and think about how the test is related to Newton's laws and to what is described as mass.

Is having mass the only way to distinguish one object from another? If so, why is that so? If not, why not?

Newton's first law is important to remember when studying the mechanical motion of an object or objects. However, we cannot empirically prove whether the first law holds or not. Think about a reason or reasons for that using the universal gravitational law and Newton's third law. You can think of one using the uncertainty principle.

Search for a definition for the gravitational mass and the inertial mass. Think about why we need the two different types of mass and how they are related to each other.

Day 16
Newton's first law

"When there is a single object present in
a system, and if that is the only object in
the system, then there is no acceleration."

If no one is bothering you, will you change your behavior or not?

From the previous lesson, we now know that the magnitude of the gravitational interaction between two objects can be quantified by their gravitational mass and the distance between them. It is great to know that, but is there something more to it? Is the law going to be the only one that we need to know in order to quantify mechanical motions in general? No, there are more. We have Newton's laws. The three laws that Newton proposed not only enable us to understand motions in kinematics but also allow us to introduce one more important quantity to be part of physics: mass. If the universal gravitational law allows us to calculate the size of the force, Newton's laws are going to tell us more about how the kinematics that we studied before are related to dynamics, which is mainly about understanding the size of the force. Amazingly, there happen to be three Newton's laws, so let us go over them, starting with the first law: the principle of inertia.

Imagine that you live on a planet that is much smaller than Earth but made of the same material that Earth is made of. That means that the size of the planet is smaller but has the same density as Earth. With all probability, the gravitational pull by the planet is smaller than that by Earth. From what? Yes, it is from the universal gravitational law. Now, imagine that you are going to throw something up in the air when you visit the planet, hypothetically speaking. Question: what is going to happen? Well, to make a long story short, the ball is going to move a lot farther than if you do the same on Earth; if you throw a ball in the air and it is displaced by 1 mile in the horizontal direction on the surface of Earth, you can do the same on the planet and the ball is going to be displaced by, for instance, 5 miles; it is going to be longer. The point is this: the object is going to be displaced farther down from where it was thrown, when everything is the same but the planet on which you throw an object up in the air. Figure 16.1 is drawn to illustrate the point.

Now, let us imagine that the planet is going to get smaller and smaller, to the size of a pen. Assume that the density stays the same as before. It is a hypothetical situation, but let us just take it as it is for a moment. There is a very small planet with a density the same as Earth. Can you guess what is going to happen when you throw a ball while in the "pen size" planet? Yes, the object is going to be displaced way farther up in the air, and potentially into space, escaping the gravitational pull of the planet—maybe 100,000 miles away from where you were, or it could be 1,000,000 miles away, depending on how strong the gravitational pull is. The point is this: the smaller the gravitational pull by the planet, the farther the ball is going to travel up in the air. Why is this important? It is because it is going to lead us to imagine and think about one very important hypothetical situation to help us better understand Newton's laws.

Now let us imagine that you throw a ball on a planet, and when the ball leaves your hand, the planet, you, and everything else in this universe disappear. That means that all we have is the ball with a certain velocity from throwing it, and that is all we have. Question: can you guess what is going to happen to the ball? So, the critical assumption is that there is no other entity "pulling" the ball via gravitational interaction in space. Answer: The ball is going to keep moving with the initial velocity, forever.

If there is an object in space and it is in motion and that is the only object in our universe, the object is going to keep moving with the same velocity, forever.

It is going to move with the same velocity as when it leaves your hand. It is going to keep that velocity from beginning to end. No change is going to be observed in the velocity of the ball, and that, in essence, is what Newton's first law is about: the principle of inertia. An object has what is called "inertia," and that does not change. When there is no net force acting on an object or objects in gravitational interaction, there is going to be no change in the velocity of the object, or the kinematical quantity associated with the object. That is what the first law is about. Figure 16.2 might help you to understand the point.

When there is no net force acting on an object or objects in gravitational interaction, there is going to be no change in the velocity of the object, or the kinematical quantity associated with the object.

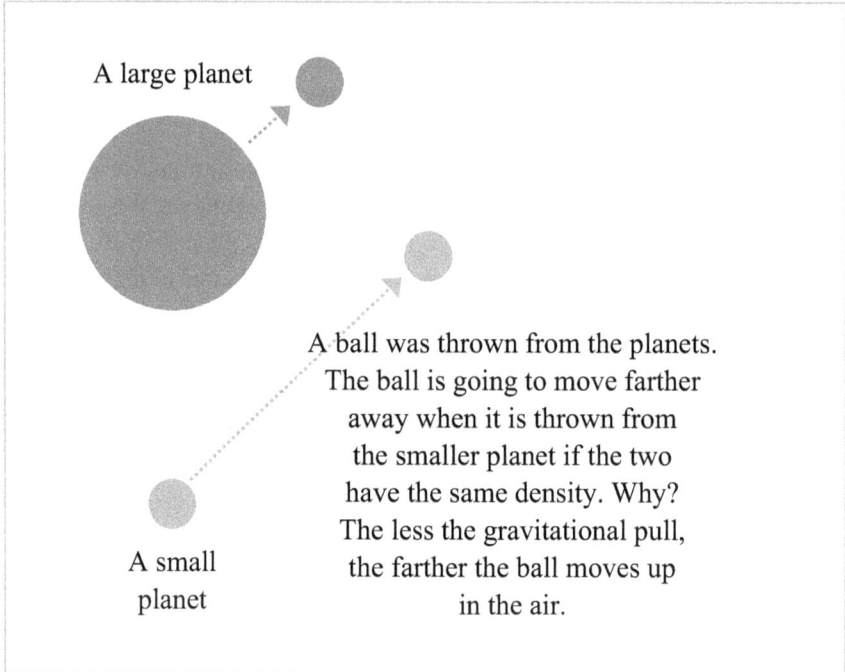

A large planet

A ball was thrown from the planets. The ball is going to move farther away when it is thrown from the smaller planet if the two have the same density. Why? The less the gravitational pull, the farther the ball moves up in the air.

A small planet

Figure 16.1: Imagine that there are two different types of planets. The one on top is a larger planet than the one on the bottom. Assuming that the planets are made of the same material, the objects that were thrown straight up from the planets are going to travel a different distance because of the different magnitude of the gravitational force on the ball. Guess what is going to happen when the planet is even smaller than the planet at the bottom of the figure. The object is going to travel a much longer distance. Then, guess what will happen if the ball is thrown and the planet and the person who threw it disappear. Yes, the ball is going to keep moving forever. The kinematics associated with the object with no net force acting on it is not going to change. That is what Newton's first law is about in short; there is no change in velocity associated with the object if there is no net applied force acting on the object.

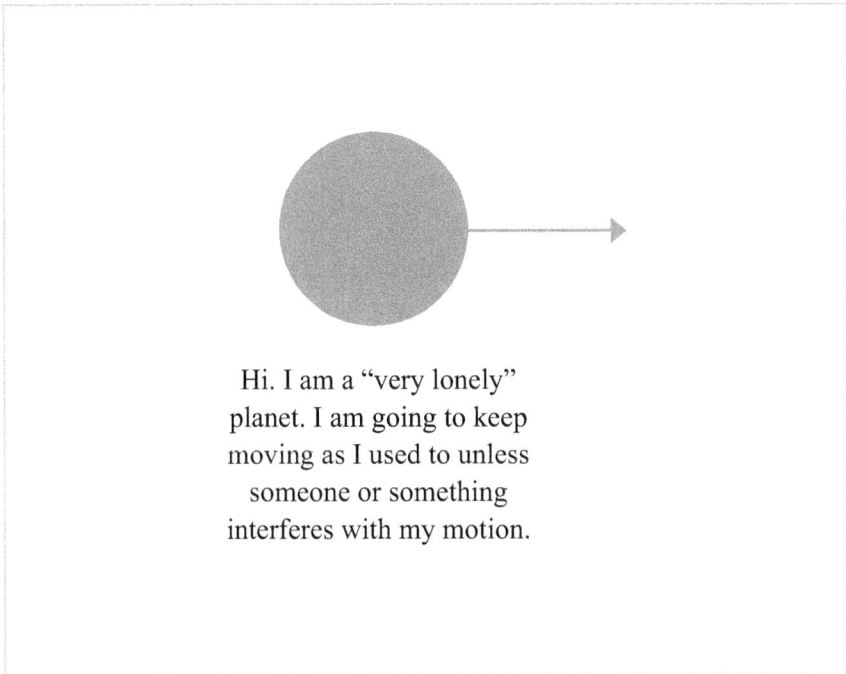

Hi. I am a "very lonely"
planet. I am going to keep
moving as I used to unless
someone or something
interferes with my motion.

Figure 16.2: If the object in gray is the only object present in our universe and it somehow happens to move with a certain velocity to begin with, then it is going to move with that velocity forever. It may be hard to imagine such a case, but that is what Newton's first law is about. Objects abhor changes in their state.

Remember: Understanding Newton's first law is important in that it is going to be the basis for our understanding the rest of Newton's laws.

Problems:

Can you realize a state where only one object is present in a system? If so, why it is so? If not, why not? Hint: you may want to think about the uncertainty principle in physics.

Can Newton's first law be considered part of Newton's second law? If so, why? If not, then why not?

Day 17
Newton's second law

"We have two objects moving at the same
acceleration but two different masses.
How do we differentiate the two?"

Imagine that you are going to run 100 m straight, and you happen to accelerate by 5 km per hour squared. You are going to keep accelerating from the beginning to the end with the same rate of acceleration. We know that it is almost impossible to keep such a motion practically speaking, but let us just say that you are able to do so.

Now, imagine that there is a car moving the same 100 m with the same rate of the acceleration, just like your acceleration for a while. So, there are two objects with different masses, and they accelerate at the same rate and are moving the same distance. Here comes a question: if someone was standing still and collided with you, or collided with the car, after you and the car accelerated while being displaced by 100 m, then which case is going to have more "impact" on the person standing still on the ground? Is it going to be you or the car? The answer: It is going to be the car. Well, we assume that the mass of the car is heavier than you. If the mass of the car is heavier than you, then the car is going to have more impact on the person standing. The heavier the object, the more impact it is going to make. You may ask the following question also: how do we quantify such an impact? Answer: That is exactly where we introduce Newton's second law: the total net force on an object or a set of objects is proportional to the size of the mass of the object and the size of the acceleration of the object. The force as a result of the "impact" is going to allow us to distinguish which entity is going to have more impact on the person standing still in the illustrated case above. Figure 17.1 might help you understand the point better.

The total net force on the object is
proportional to the mass of the object
and the magnitude of the acceleration
of the object.

Or, using a mathematical expression:

$$\text{Force} = \text{Mass} \times \text{Acceleration}$$

In other words, if you have two objects that look the same and they are accelerating with the same rate but with a different mass, you are not going to be able to identify which one is which while the two objects are in motion without having an interaction with another object. If they are hit by another object or set of objects, then we will be able to identify which one is which by seeing the "impact" on the other objects. The difference is in the mass, and that's what classical mechanics is all about.

Here is another important point: note that the force that we are discussing in Newton's second law is just a "force." It is a force that we understand as the mass times the acceleration in general. What is my point? The force that is being discussed in Newton's laws does not have to be a specific type of force, such as the gravitational interaction. It could be any type of force, including the gravity that we have discovered or are going to discover in the future. The mechanical motion in physics can be induced by any force, not necessarily a gravitational force. Once the size of the interaction is quantified by that, such as the universal law in gravitation, where the degree of gravitational interaction is quantified, then we set that equal to Newton's second law to see the acceleration associated with an object. The law lets us analyze the motion from the dynamics that runs behind. Remember that.

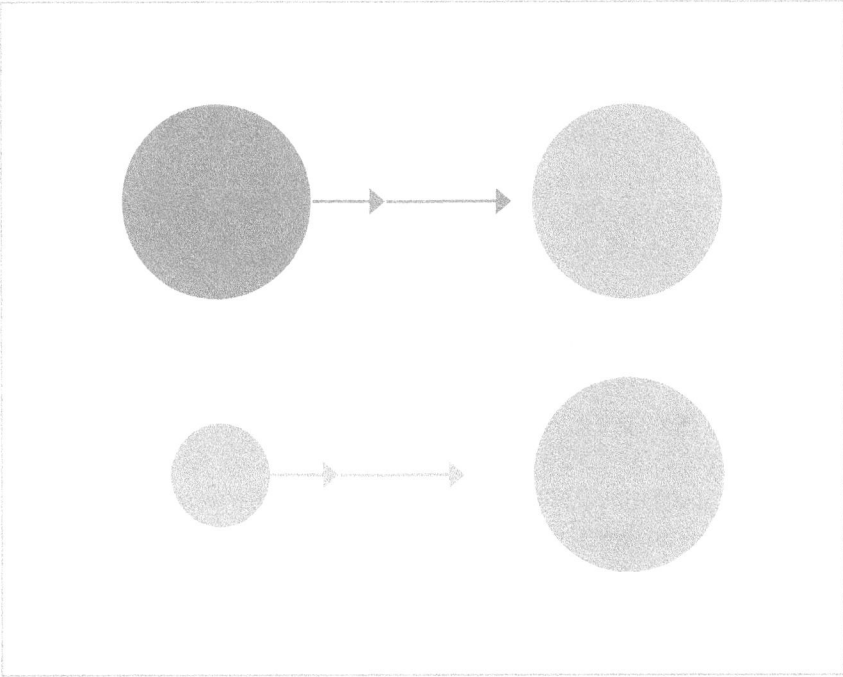

Figure 17.1: Imagine that the sphere on the right-hand side is at rest to begin with and that the one on the left-hand side moves to the ones at rest. Both the light gray and the gray objects have the same density, and they are going to accelerate to the same value just before they are going to collide with the grey sphere. Question: they have the same rate of acceleration but with different masses. Which one is going to distort the shape of the grey sphere more when they collide with each other? Why does the size of their mass play an important role in this scenario? If it does not, why not?

Remember: Newton's laws are not only important but also very useful. What we have learned in kinematics is related to what we are learning here in the dynamics section.

Problems:

Imagine that we have an object on a planet. Gravity is pulling the object down with 100 Newtons of force where 50 Newtons of air resistance is present. Calculate the acceleration associated with the object. Assume that the mass of the object is 10 kg. Calculate the gravitational acceleration associated with a person on the surface of the Earth. You need to use the universal law in gravitation with Newton's second law in order to set up an equation.

Calculate the gravitational acceleration on a small object caused by the presence of the Earth. Describe why the acceleration is not going to work for an object whose mass or size is comparable to that of Earth. Hint: go over the lesson on the two Earths.

Jae made a hole that goes all the way down to the center of the Earth, and he is going to travel to the center. Calculate the size of the acceleration on Jae caused by the presence of the Earth when Jae is about 25 percent of the radius of the Earth down.

Day 18
Newton's third law

"Force comes in a pair, always."

There is no such force as a single force. An action must be accompanied by a reaction.

Let us go back to you throwing a ball up in the air again. Imagine that you are going to throw the ball as far as you can. In fact, let us assume that you did the best you can do. You are going to throw the ball as hard as you can.

Question: when you throw the ball forward as hard as you can, what is going to happen to you? We are talking about the motion associated with yourself. Is something going to happen in terms of the physical quantity associated with yourself?

Answer: Yes, something is going to happen to you. You are going to be displaced back as the ball moves forward. You or anyone in the universe cannot throw something up in the air without being displaced one way or the other. You must be displaced in order to displace something else in space. It is always mutual. You may think you can just throw a ball while you are standing still in a fixed position, but the fact of the matter is that if a ball is thrown by you up into the air, then you are going to be displaced backwards, always. If you truly are not displaced when throwing the ball, then think about what needs to be displaced. You may want to think about the very first lesson in this book.

Let us go over one more case. Let us say that someone positions a ball 100 miles away from the ground surface. That means that the ball is way up in the air, and it is going to start falling toward the surface of the Earth because of a gravitational interaction. Question: what is going to happen to the Earth while the ball is moving down? Yes, the Earth is going to move up, although the displacement of the Earth is so tiny, thus almost negligible on a macroscopic scale.

The point is that there is no single object that could realize being in a motion on its own. That is what the principle of action and reaction is about in classical mechanics. According to Newton's third law, an action must be accompanied by a reaction, always. So, when an object gains velocity, or realizes a change in its velocity, there must be something or someone giving the object that velocity, and that something or someone is going to be displaced backward in their reaction. So, things are going to get balanced out.

Force comes in a pair, always.

An action comes always with a reaction. That is what Newton's third law is about. Figure 18.1 illustrate the point well.

> When there is an action, there must be
> a reaction.

Do note that the balance in the action and reaction is realized in terms of "force" in dynamics. This is something important to remember as you learn more. You are going to study and read about other dynamical quantities such as energy and momentum later and in other literature, but the fact of the matter is that the mechanical motion where the three Newton's three laws can be manifested is always via force. You cannot do that with the energy and momentum. That is where Newton's laws gain their importance, and that is one of the reasons that we study Newton's laws to understand mechanical motions.

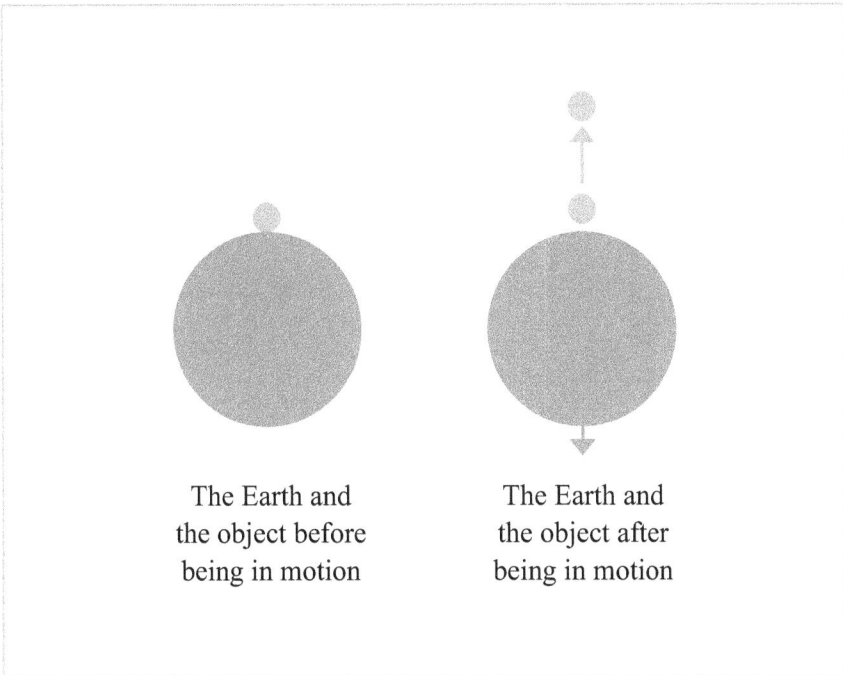

The Earth and
the object before
being in motion

The Earth and
the object after
being in motion

Figure 18.1: Imagine that the light gray object represents a small object and you throw it up in the air in the vertical direction while you firmly stay on the surface. Is the object the only one being displaced in the system? No, it is not. As the object is being displaced up, the Earth is going to be displaced down, although the size of the displacement of Earth is going to be much smaller than that of the small object that you throw up in the air. Again, force always comes in a pair. For that reason, if a change occurred to an object within a system, another change should occur to another object within the system. When that happens, the size of the force that acts on the object is the same size but going in the opposite direction. That is what the principle of action and reaction is about in essence.

Remember: Understanding the essence of Newton's laws is a core part of classical mechanics. Without understanding Newton's three laws in depth, you are not going to understand the rest of the materials covered in this book or in your course. For that reason, I am going to highly encourage you to spend some time getting familiar with the three laws by going over some practice questions.

Problems:

Find a case where the principle of action and reaction does not matter much. Hint: go over Newton's first law and think about a case.

Find definitions of inertia in classical mechanics and write a paragraph about how it is related to an object moving with the same velocity unless there is an outside force acting on the object.

The size of the gravitational force caused by Earth on us is very large, so, in principle, we need to be pulled down to the inside of the Earth. However, as we all know, we are joyfully staying on the surface of the Earth. Why is that?

Think about how velocity is related to acceleration in physics and think about where Newton's laws fit into the relationship. In other words, think of a physical quantity that is closely related to the force, and show how that force is related with respect to time.

Day 19
Weight

"Weight measures force, but mass does
not. Mass is the same no matter how
strong or weak the gravitational interaction is. Mass does not change in classical
mechanics. In short, weight depends on
the size of the force, whereas mass is
invariant."

One of most typical questions that physics instructors can ask as a part of
their test questions is the following: describe a difference or differences
between mass and weight.

Why? Because so many students have issues with clearly understanding the difference between the two quantities when taking physical
science or college physics. I have seen that throughout many years of
teaching physics, and I do believe that many other instructors have had a
similar experience when it comes to the issue. So let us address the issue
here one more time. We are not studying anything new in this lesson but
are trying to shed a new ray of light on the issue.

Weight measures the size of the force acting on an object. Again,
weight is a force whereas mass is not.

What is the main point? It represents mass scaled by the acceleration, not
just the mass. So, when it is about the weight because of the presence of
the Earth, then it is the same as the size of the gravitational pull by Earth.
Given that the size of the acceleration is commonly taken as a constant,
many students think that mass is the same as weight. Well, they are
not the same.

Weight is a force. In other words, the physical unit associated with
weight is force, not mass. Weight changes depending on the size of the
gravitational interaction, whereas mass is invariant.

Let us go over a simple case, which is a common mistake made by many
high school and college students. The Earth pulls an object near to the

surface of the Earth because of their gravitational interaction. Again, the size of the gravitational pull is what we define as weight, which is just the degree of their gravitational pull, and the size changes depending on with which the gravitational interaction takes places. For instance, for someone staying near the surface of Mars, the size of what we define as weight, the gravitational pull by Mars, is going to be quite different from that by Earth. When the weight of the object happens to be 100 Newtons on the Earth, the same object has about 40 Newtons of weight when it is placed near the surface of Mars. Why? The degree of the gravitational pull by Earth and by Mars are different, almost by a factor of 2. That degree that is due to Mars is only about 40 percent of that due to Earth. It is as simple as that. Weight indicates the amount of force, not the mass, which is invariant no matter how strong or weak an object is that is going to interact with others. Remember that weight does depends on which reference object we are dealing with. Figure 19.1 was created to illustrate the point. Keep in mind that the mass and the weight have different physical units; they have different dimensions. For that reason, they cannot be directly compared. The unit for the force is Newton and for the mass is kilogram in the standardized unit system.

What about mass then? In classical mechanics, a convenient way to understand what mass is about is to start from Newton's laws. As being covered in dynamics, mass is something that is introduced in order to distinguish an object from the others. However, if someone asks you what mass is, or what "inertial" mass is, then it becomes a different story, and we may need to understand the difference between inertial and gravitational mass. If someone asks you how mass is created, that is another different story. For now, let us just understand that "mass" is invariant, whereas weight depends on the size of the gravitational pull.

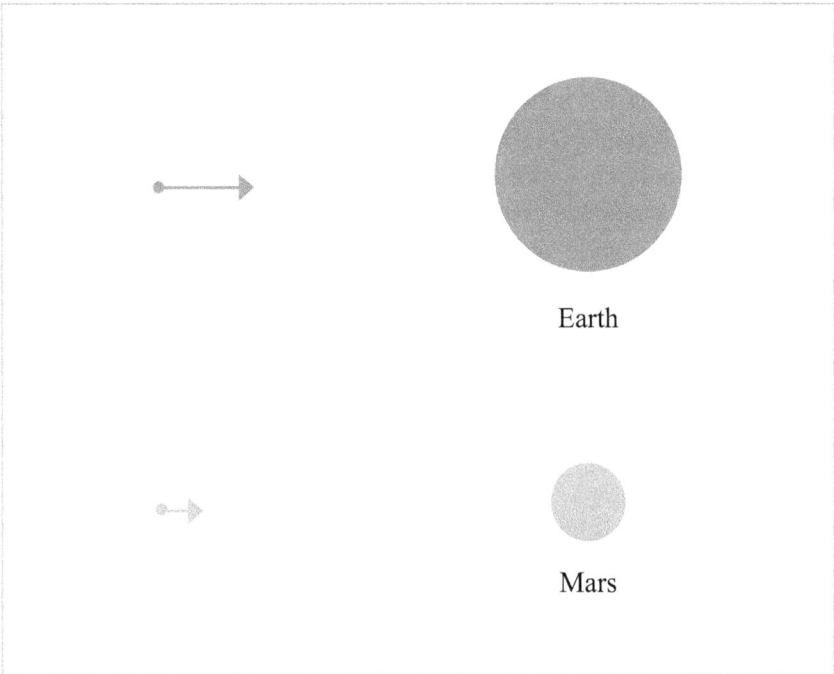

Figure 19.1: Weight indicates the size of the gravitational pull in most cases. Weight is about force, whereas mass is about the amount of matter contained in an object. When the object is gravitationally pulled by the Earth, and the size of the gravitational force is 100 Newtons; the same is going to be about 40 Newtons when it is measured with respect to Mars. That is because the mass of Mars is smaller, thus the size of the pull is going to be smaller.

Remember: understanding the major difference between weight and mass when taking physics in high school and college is highly important. Why? Because it indicates whether you have some understanding on what gravitational acceleration is really about. For that reason, if you do not think you understand the differences clearly after reading this lesson, I encourage you to take some time to go over some of the previous lessons again until you are sure that you understand the difference. Again, to make a long story short, weight is a force, and it is not the same as mass.

Problems:

Think about why it is hard to define mass in words. Is there a way to directly measure mass or not? Read some literature that covers some different types of masses and think about which type of "mass" is what we usually refer to when we study undergraduate classical mechanics.

Estimate the ratio of weight and mass for small objects close to the surface of the Earth. The word "small" means that the mass of the object is much smaller than that of the Earth.

Calculate the weight of a 10 kg object close to the ground surface caused by the presence of the Earth. Do the same when it is measured on Mars. Is the weight associated with an object the same no matter where the object is located?

If mass is the total amount contained in an object, then an interesting question follows: How do we measure the mass associated with a single object? Can we ever measure the size of a "true" mass that is associated with an object? Go over some books or articles that address this issue.

Day 20
Two Earths

Jae: Imagine that we have two Earths that
are the same—our Earth and another that
we just discovered last week.
Adam: Really?
Jae: What's worse is that they are about
to collide. How do we calculate the
size of the acceleration associated with
the Earths?

Imagine that we have two of the same Earths. I know that is a hypothetical situation, but I have decided to go over the case briefly because
I strongly believe that it will help you understand Newton's three
laws better with the concept of the center of mass and relativity being
introduced.

Let us start with a figure. In Figure 20.1, you see that the two Earths
are about to collide. Here comes an interesting question: what is the distance between the two in classical mechanics just before they are going
to hit each other? Answer: We do not measure the distance with respect to
the surface but with respect to their center of mass. When we do so, the
distance is going to be about twice the radius of Earth. Why? We measure the distance with respect to the center of mass, which is located at
the center of Earth. We assume that Earth is a perfectly spherical object.
So, you simply double the radius of one of the two Earths as the distance
between them. Do note that you do the same when you measure the distance with respect to a person standing on the surface. However, since
the size of a human being is much smaller than the size of Earth, the distance from Earth to you is almost the same as the radius of Earth. In other
words, we can ignore the size of a human being when it comes down to
comparing it to the radius of Earth.

Let us focus back on the gravitational acceleration. Again, we have
two of the same Earths, and they are about to collide. The distance
between the two is twice as large as the radius of our Earth. So, the
acceleration is going to be reduced by a factor of 4, when considering
the distance part only. If the acceleration caused by our Earth on us is

10 m per sec squared, an approximated value, then that caused by one Earth on the other Earth is going to be about 2.5 m per sec squared, just by doubling the distance. Is that the end of the story? Well, I may need to point out one more thing. Going back to the very first lesson, we need to set a reference when reporting a physical quantity. The size is going to be 2.5 m per sec squared when the acceleration of one Earth is measured by someone very far away from the system of Earth. This is an important point to understand. In other words, when someone whose mechanical state is not going to get affected by the presence of the two Earths measures the acceleration, approximately speaking, then it is going to be 2.5 m per sec squared. Again, someone who is very far away from the location of Earth measures the acceleration, and it is going to be about 2.5 m per sec squared. Question: what about the acceleration of the Earths with respect to each other? With respect to one Earth, what is the acceleration of the other one?

Answer: You just double the size of the acceleration. Our Earth accelerates the same as the other one but in the opposite direction. Let us go back and review the case that was covered in the very first lesson: you drive your car 100 miles per hour on a highway, and a police car is driving with the same speed but in the opposite direction, with respect to someone standing still on the ground. You are going to think that the police car is approaching you going 200 miles per hour; the quantities are relative depending on where we set a reference. On the same token, if the reference is our Earth when measuring the size of the acceleration associated with the twin, we simply need to double the size. In other words, the acceleration of our Earth with respect to its twin is 5 m per sec squared. The same goes for the twin Earth. Figure 20.2 is all about the story.

In short, we can ignore the mass and the dimension associated with a human being when calculating the size of the acceleration of the human caused by the Earth, practically speaking, and we can do so since the mass and the size associated with a human being are much smaller than those of the Earth. In other words, when the mass and the size of an object are not negligible, we need to consider them when working on problems. There we need to introduce the concept of the center of mass. For instance, if we have our Earth and its twin, and they are about to collide, we need to consider the size of the twin and the mass of the twin when calculating the acceleration. For sure, we need to set a reference point properly too. What happens if the mass and the size of the twin is different from that of our Earth? Then things are going to get more complicated, and I will leave that up to you.

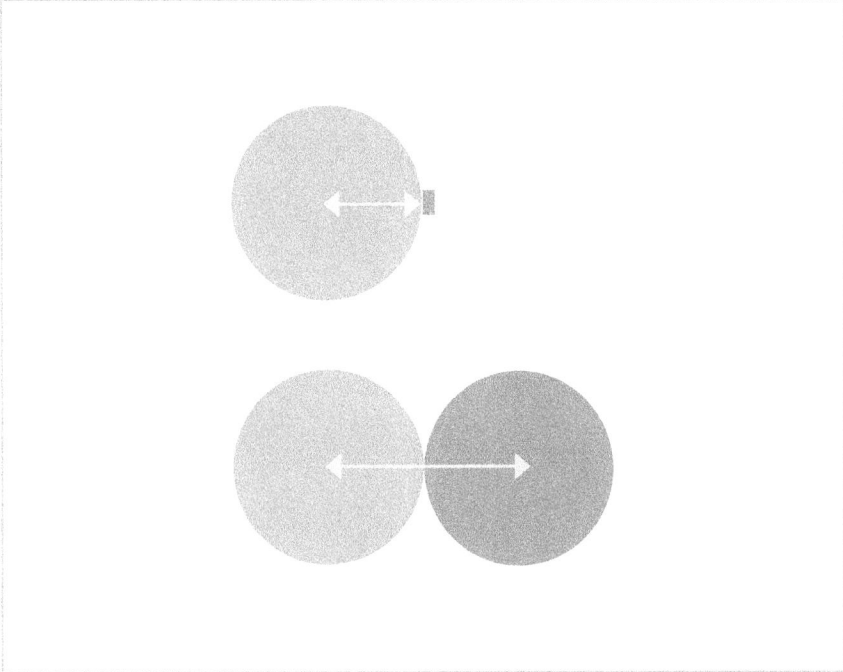

Figure 20.1: This shows the two cases that are relevant when calculating the size of the gravitational acceleration on us that is due to the Earth on the top and that due to the other Earth on the bottom. The sphere in grey is our Earth, and the small box drawn right next to the grey sphere represents an object in the scale of our human. When calculating the distance between the human and Earth, do we need to consider the size of our human in practice? We may not. The human is much smaller than Earth, so it is okay to treat the human as a test particle in this case. However, that may not be the case when we calculate the acceleration of the other Earth caused by our Earth. The dimension is comparable, and we need to consider the size of the Earths. When we do so, the distance between the two is not the same as the radius of the Earth but twice of that, as illustrated on the bottom portion of the figure. Question: when the distance doubles, what happens to the size of the gravitational interaction? Yes, it is reduced by a factor of 4. Think about why that is so.

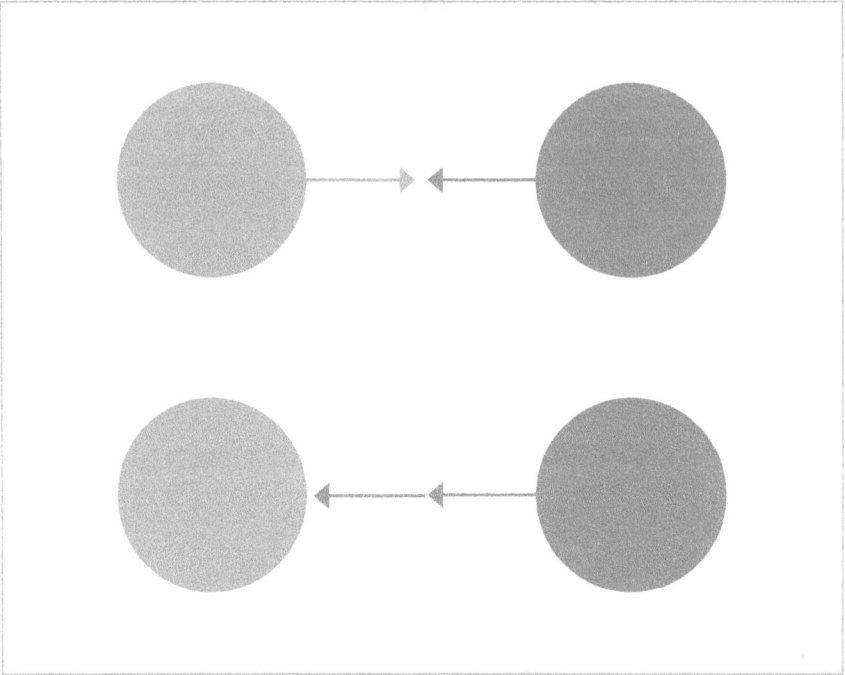

Figure 20.2: This illustrates the point of quantities being relative in physics one more time. On the top, we have two of the same Earths, our Earth in the light gray and a twin in the dark gray. When someone far away from the system of Earth observes the Earths pulling each other and accelerating toward each other, the person is going to measure the acceleration of our Earth to be about 2.5 m per sec squared when the two Earths are about to collide. On the other hand, with respect to our Earth, the acceleration doubles. Just like the way the relative velocity is illustrated with my car moving 100 miles per hour and a police car moving 100 miles per hour on the opposite side of the lane, and my thinking that the police car is approaching me at 200 miles per hour, if we set our Earth to be the reference, which is represented as a sphere in grey on the bottom, then the acceleration of the twin is 5 m per sec squared, not 2.5 m per sec squared. Again, we need to think about the size of the objects in the system, and we need to think about how we set a reference when reporting the size of a quantity such as an acceleration. Things are relative. Remember the very first lesson we had.

Problems:

Imagine that our Earth is about to collide with another planet, a planet
that is the same size as Earth but 10 times heavier than Earth. Calculate
the gravitational acceleration of our Earth with respect to the planet
and then that of the planet with respect to Earth. Assume that the
densities of Earth and the planet are the same.

Day 21
Quantities in dynamics

"Length and time are two useful quantities when describing kinematical motion
associated with a single object. As our
introducing mass as one more useful
quantity, we can describe dynamics
where we have more than one object in
the system. Think about combining what
we have studied with mass."

Question: you have studied all about kinematics. Now with mass
as an extra quantity, what do you do with all that you have studied?
Answer: You combine the two.

Let us think about a free-fall of an object briefly. If you hold an
object and release it, then the object is going to fall vertically. If you
repeat this, dropping an object many times, the result is going to be, with
all probability, almost the same; the object is going to fall to the surface
of the Earth. It is as simple as that.

Now, here comes an interesting question: why does the object move
down toward the surface? Have you thought about that in depth before?
Some of you who have read classical mechanics before or have pondered
about an apple falling from a tree might already know the answer. Newton thought about that too. Why is this important? It is important because
this is one example where dynamics in classical mechanics is well illustrated; when we have more than one object within a system, they are
going to interact with each other. There could be different types of interactions in physics. In classical mechanics, the gravitational interaction is
what we mainly focus on. It is one of many different types. Understanding and thinking about an object falling to the surface of the Earth is one
of the effects of having such interaction in space.

Let us go back a little bit. In kinematics, what we have studied was
about the motion associated with an object or objects, but without understanding the cause for the motion; we analyzed the mechanical motion
with a critical assumption. Again, the objects somehow gained their kinematical quantities from somewhere at some point, and we did not raise a
question regarding where the physical quantities came from or by which

interaction mechanism such quantities are gained. The story is going to be a bit different here. In the dynamics part of classical mechanics, we are going to raise all the questions together and think hard about the "why" part. In other words, what we study here in the dynamics lessons is about the "cause" part; why has such an object gained the motion and by how much? There, we are going to use dynamical quantities, such as force and momentum and energy. Does that ring a bell? Yes, if you read the cause-and-effect lesson.

In terms of the cause and effect in physics, the quantities in dynamics are something that leads us to understand the "cause" part. By doing what? By introducing "mass" as an additional physical quantity. When we do so, objects interact with each other gravitationally, so the fun of analyzing a system using dynamical quantities begins. How? In terms of physical quantities in classical mechanics, we just need to combine the mass with other quantities that we have studied before. So, in most classical mechanics courses, there is a reason for instructors to cover the kinematics part first.

So, think of it this way: you studied all kinematical quantities when analyzing the motion of a single object in a system of our interest; now we want to extend the case that we studied there by introducing one more object into the system. For instance, in kinematics, we mainly played with displacement and velocity and acceleration as the three quantities. When we combine acceleration with mass, that is exactly what force is about. Following Newton's laws, inertial force is proportional to mass and acceleration. On the same token, when we combine velocity with mass, we end up with what we are going to study as "momentum," a crucial quantity that we play with in a collision process, and it happened to be proportional to mass and velocity.

You may ask about displacement too. Interestingly, when we combine force with displacement, we end up with "work done" to or by the system. We are going to study all the quantities later, and this is something to do with what we study as "energy." These are quantities that we can use to better understand the dynamics associated with a system where we have more than one object. Again, if we have a single object in a system, we do not need to introduce the dynamical quantities because there is no other object to be compared with. However, whenever we have more than one object in a system, we need to think about the causes of their motions, and we need to introduce dynamics quantities. We just need to be more realistic. Welcome to the dynamics part of classical mechanics.

Remember: in dynamics, we do not deal with a single object anymore but with multiple objects in a system. There you combine the
quantities that you have studied in kinematics with mass, and that is
where the useful quantities, such as force, momentum, and energy, get
introduced.

Problems:

Briefly describe why we do not need to know the mass associated with a
single object if that is the only object in a system. Hint: does it matter?

A change in momentum as a function of time happens to be force. Prove
it mathematically.

Think about a few differences between the quantities that we studied in
kinematics and in dynamics. What is the major difference between the
quantities in the two groups?

Day 22
Force

"This is what we really need and what
turned out to be highly useful when
describing gravitational interaction in
classical mechanics."

Imagine that we have two objects in a system. We have only the two objects in the system, and they happen to be about 1 m away from each other. Question: among force, momentum, and energy, which one helps us most when trying to analyze and understand the motion associated with the two objects because of their gravitational interaction, particularly when they are at a distance? Or, phrasing it differently, which quantity do we need more than the other dynamical quantities to begin analyzing the mechanical motion associated with the objects in the system? Answer: Yes, force is going to be the one. Why? Because force is where Newton's laws are governed and manifested, so we can utilize the laws in order to analyze the motion.

Force is a dynamical quantity through
which Newton's laws are manifested. It is
neither energy nor momentum. We need
to know about force in order to utilize
Newton's laws.

Is that really so? Well, let us briefly go over a simple case. For the sake of keeping matters simple, we assume that the coupling constant, or coupling strength, is unity, so we can ignore that. We also assume that one of the objects is 1 kg and another one is 2 kg, and they somehow happen to be at rest when they are 1 m apart from each other. Question: how do we calculate the acceleration the two objects begin with, the initial rate of the acceleration? First, we calculate the size of the gravitational force between them using the universal gravitational law and then divide the force by the mass of the object. For instance, if we want to estimate the acceleration of the 2 kg object, we simply divide the force by 2, so the acceleration of the 2 kg object is going to be 1 m per second squared. Then we can calculate the velocity of the object from the acceleration and the time elapsed during which the acceleration was taking place. In other words, we do the following: we calculate the size of the total force. Set it equal to the mass of the object multiplied by the acceleration. Then solve for acceleration.

In summary,

Total external force acting on an object = Mass × Acceleration

In short, when two objects are at rest and at a distance, then we can calculate the acceleration of the objects using force, taking Newton's laws into account.

Now, let us think about the other two dynamical quantities for a moment. Can the information regarding momentum or energy associated with the objects be utilized to calculate the acceleration associated with the object or objects? No, it cannot, not directly at least, particularly when the distance between two objects is given or measured. A critical point here is that the force on each object is going to be the same, following the principle of action and reaction. It is neither momentum nor energy by which the size of the acceleration associated with the two objects can be easily calculated; momentum is something that we can calculate once we calculate the velocity from the acceleration. Energy is something we can certainly utilize but, again, in order to apply the principle of action and reaction. Therefore, the bottom line is that we do need force as a physical quantity when analyzing motions in dynamics. It is very important, especially when the two objects are at a distance. You might ask if momentum is also very important. Yes, it is very important, especially when the two objects that we are interested in are about to collide. There, the conservation of momentum is going to be a key player to be utilized.

Does that mean that force is a more important quantity than momentum and energy? No, it does not. They are all important when understanding mechanical motion, and more details will be covered in the momentum and energy part. In short, both momentum and energy are very important when describing the collision process in dynamics, and energy is particularly important when trying to understand the motion associated with objects with dimensions, objects with a certain shape in space. In the end, all three dynamical quantities are important, but when it comes down to practice, it depends on what we are to analyze and understand. Among them, force is where the principle of action and reaction and the other two of Newton's laws can be applied, so there it gains its usefulness. Remember that.

Remember: force is where Newton's laws can be used. It is neither momentum nor energy. That is one of the reasons why force is covered along with Newton's laws in classical mechanics. Force is where the laws can be applied.

Problems:

Describe what is going to happen if energy is where the principle of action and reaction can be applied instead of force.

Calculate the size of the gravitational acceleration on a small object caused by the presence of the Earth, using the mathematical expression covered in this lesson.

Think about whether we can quantitatively measure force as we do displacement and velocity in space. Think about what we need to do in order to measure displacement in space and think about whether we can do the same for force.

After you read the lessons that cover momentum and energy in this book, describe the advantages and disadvantages of the quantity "force" compared to "momentum" and "energy." Describe why it is so helpful for us to describe what happens in a system when we use all three instead of just one or two.

Day 23
Tension and friction and normal force

"Objects have dimensions in space, so we
need to introduce quantities to describe
the kinematics and dynamics of objects
when they are in contact with each other."

Question: can you easily think of an object or objects that are not in contact with another one?

Let us discuss reality for a moment; almost all objects that we deal with on a daily basis are in contact with something else. We are going to be a bit practical on this issue. Imagine how many objects that you see today that are not in contact with others. It is hard to find one that is not in contact with others. Point: we need to understand the physics that describes states associated with objects in contact with others. Keep in mind that we want to be as specific as possible when describing a state associated with objects. That is one of the reasons for our coming up with velocity and acceleration back in the kinematics lessons. That also is why quantities such as frictional force, normal force, and tension get our attention, and here it comes: let us study them one by one.

When two objects are in contact with each other, we need to think of their gravitational interaction differently from when they are at a distance. Why? In practice, objects have dimension in space. They occupy space, so we need to introduce different types of forces that allow us to describe their states. Hint: if you are interested, read the lesson on the two Earths. You will clearly see why this is needed.

Let us start with tension. Imagine that there is an object on the ground and you want to pull the object up into the air on your own. What can you do? One way to do so is to attach it to a rope and pull the rope up instead of directly lifting the object up. What happens to the state associated with the object then? The force that you applied to the rope is going to be transferred "along" the rope, a flexible medium, to the object so that you can pull the object up in the air. In essence, you deliver the force indirectly to the object. Force gets transferred. There is a mediator. Do you see the advantages of that?

Tension, by definition, is a force that is transferred along a medium or mediums. As illustrated above, it is for our convenience, just like all the quantities that have come up so far. You do not need to handle heavy objects on your own by grabbing them and lifting them up; you let a medium to do the job directly. Again, as the definition goes:

Tension is a force that is transferred along a medium.

The force that you are applying to the rope is going to be transferred to the object along the rope so that the object is going to stay up in the air. How? The force that you are applying to the rope is balanced by the gravitational force on the object from the Earth and Newton's second law. In other words, instead of holding the object directly; you can hold the object indirectly using a rope, a medium where the force you are applying is transferred to the object. Figure 23.1 illustrates the point. If it is a heavy object with volume, it is going to be a bit hard for you to hold it using your hand, but it is easier if it is being held by a rope.

Imagine that there is a very large 100 kg box on the ground surface, and you want to hold it and lift it up in the air. It is going to be a bit hard for a single person to hold such a heavy object up in the air and displace it to another place.

How about the case when we have two or more people to do the same task? Obviously, it is going to be easier to accomplish the job. For instance, we can hold the object using ropes—two ropes. Then each person could deliver the same amount of force to the rope, so, in this case, if the weight of the object is 1,000 Newtons, then each person needs to deliver 500 Newtons to the rope. If one person is to hold the object, the person needs to provide 1,000 Newtons to the rope.

In addition, we can change the direction to which you apply force. Again, we are going to hold the object using a rope, but this time we are going to have a pulley and let it change the direction of where the force is being transferred; there is a pulley fixed up at a distance. As a person holding the rope is pulling the rope down, then the object is going to move up since the pulley is where the direction of the applied force gets changed. The rope as a medium is not a rigid body; it has a large degree of elasticity, so we can take advantage of changing direction of the applied force to the object. Figure 23.2 might help you better understand the point; the direction associated with the applying force is not the same as that of the object in motion. It does not have to be. Or you can change it as you want to, depending on how you set up the pulleys. Just remember:

> Tension is a force along a medium or
> mediums, so it allows us to apply a same
> magnitude of force indirectly to an object
> or our interest and allows us to change
> the direction of the force we apply.

It is also where we can refresh our memory of the principle of action and reaction. For instance, when someone pulls the end of a rope, but the rope happens to be at rest, that means that a force of the same size but in the opposite direction is being applied to the other end of it. Keep in mind that force is where Newton's laws can be applied and utilized. We are going to come back to this topic when we study conservation of energy later.

There are many other works of literature with sets of practice problems that help you understand tension. Just remember the following for now: it is a type of force that we need to consider because of objects having dimensions.

Let us move on to normal force. When we have two objects and they are in contact with each other, at some point, we may need to think about normal force. Think about the following case: imagine that you stand up on the ground. Yes, you are simply standing on it, but if you think about it, you may ask the question: how come? Following the principle of action and reaction, when Earth pulls you down, you should go to the very center of the Earth. What do you think? In principle, that is how gravitational pull should work. The Earth is a big and mighty object, so you are going to be heavily pulled in. But what is happening in reality? You stay on the ground surface instead of being pulled down to the center. How come? How come you stay on the surface? Answer: The ground surface saves us. It supports you with the same amount of force as that being applied by the Earth, but in the opposite direction.

The ground surface supports you.

This is one of the typical examples where we can think about how normal force works. It is a supporting force. As you are pulled down toward the very center of Earth, the supporting force, the normal force, of the ground surface is holding us up with the same amount of force as the gravitational force, but in the opposite direction. That way, the total net force acting on you even from the presence of Earth is going to be zero. Hint: think about Newton's second law. What is the minimum condition for you to not accelerate? Anyway, due to the ground surface holding you up, you are not accelerating with respect to the ground, so you are allowed to stay up on the ground. Does that ring a bell? Yes, this is all about understanding the principle of action and reaction. Force always comes in a pair. Do you see how everything is coming together and is related?

Force always comes in a pair. There is no single force realized in classical mechanics.

If there is a force that pulled us down, there must be a force that pushes us up, so the two forces are going to be balanced out. So, no net force is acting on us, and we can stand on the surface of Earth. That applies to all other cases. Normal force can be thought as a supporting force, and again, it is going to be a type of force that is realized when an object is in contact with another object and force and they are heading toward each other. Again, this is something that we need to worry about since objects have dimensions; they have some shape, so being "in contact" is realized. Figure 23.3 illustrates the point.

> Normal force is a supporting force. In the case of you standing on the surface of Earth, the surface supports you up to balance out the gravitational force on you. Therefore, the total net force on you from the presence of Earth is zero.

There is one more left to go. Yes, it is frictional force. This is also something that we need to worry about because of objects that appear to be in contact with each other. It is a type of force that we need to use in order to realize and enjoy our lives in practice. In order to understand what that is about, we need to understand its origin first. Let us go over that together.

Imagine that we have an inclined plane, a plane that is not parallel to the horizontal direction, and a ball is rolling down the plane hard.

What do you think the motion of the ball will be? Does the ball go straight down to the surface of the Earth? Or is there something more to it?

Let us take a look at Figure 23.4. The left side shows the typical drawing of an inclined plane and a spherical object that is staying on top of it. In general, that is how we illustrate a moving object on the plane with an angle with respect to the horizontal direction. However, let us get a more realistic picture by zooming in. We zoom in very closely. If you have a closer look at the very tiny fraction of the inclined plane, very tiny, as shown on the right side of the figure, what are you going to see? Think about taking a look at the portion that is in contact with a microscope. Yes, you are going to see all kinds of structures associated with the line. There is no such thing as a straight line. You simply cannot realize it. Remember: think about the uncertainty principle in physics. The straight line that you can think of and that you studied in mathematics is just in your mind and in a hypothetical world. There is no such thing as a perfectly straight line in physical space.

We cannot realize a perfectly straight line in physical space, and that is one of the reasons that we have to deal with friction.

On the same token, any object that has structure in space is not going to have a perfectly smooth surface. There are all different kinds of wiggly shapes that you are going to observe when you have a close look. Question: imagine that an object with a wiggly structure is in "contact" with another object with a similar feature, that is, some wiggly shapes on their surface, at a microscopic level. They appear to be in contact with each other. In other words, on a microscopic scale they are in close contact with each other, and you are trying to push one against the other. What is going to happen?

Answer: Because they don't have a perfectly smooth surface, you need to apply some force to cause the object to be in motion. Furthermore, the force that you apply to the system needs to exceed the size of the "threshold" at which the force begins to overcome the resistance the objects have to being in motion because of their wiggly shapes. That "threshold" is exactly what the size of the frictional force, or friction, is about. By definition, friction is a type of force that needs to be overcome in order for an object in contact with another to be in motion. Following the principle of inertia, which is about Newton's first law, they tend to resist the presence of an external force or forces, until the size gets bigger than that of the frictional force.

In summary, frictional force is a type of force that resists the motion of an object because of the dimensions associated with objects. In addition, the direction of the friction is always opposite to that associated with the motion. Again, friction is a type of force that resists motion. In terms of the physical quantities that we have studied, the size of frictional force is proportional to that of the normal force, the supporting force that we just covered, and to the coefficient associated with the material that the surface is made of. Every material has a different frictional coefficient; you may think of it as different types of objects that look like they have a smooth surface have different degrees of "wiggly ziggly" shapes in microscopic scale.

The size of the frictional force is proportional to the magnitude of the normal force and the frictional coefficient associated with the material that the object is moving on.

Or,

$$Friction = Normal \times Coefficient$$

There is one more thing: say, again, that there are two identical objects and, this time, we put both on the same surface, but one of them is in motion to begin with and the other one is not with respect to the ground surface. The size of the frictional force that is associated with the two is different even though they are made of the same medium and with the same structure. In other words, when an object is in motion, it takes less force to keep the object moving than before. What does that mean? We need to have two different types of frictional coefficient, one that is for static and another for kinetic. The former is what we need to consider for frictional force when objects are not in motion to begin with, the static friction, and the latter is when objects are moving.

We have covered three different types of forces that we need to deal with that are due to objects being in contact with each other, and they are the following: tension, friction, and normal force. In other words, we do not need to deal with them when we have two test particles at a distance from each other. Tension is a force that is transferred along a medium, so we can adjust the size of the applied force and the direction associated with it. Normal force is a supporting force that we need to consider when an object is directly in contact with another one. Friction force is when an object resists motion because of its wiggly shape. All three forces are applicable when the objects of our interest are in contact. Again, they are forces, so they follow what is described in Newton's laws too.

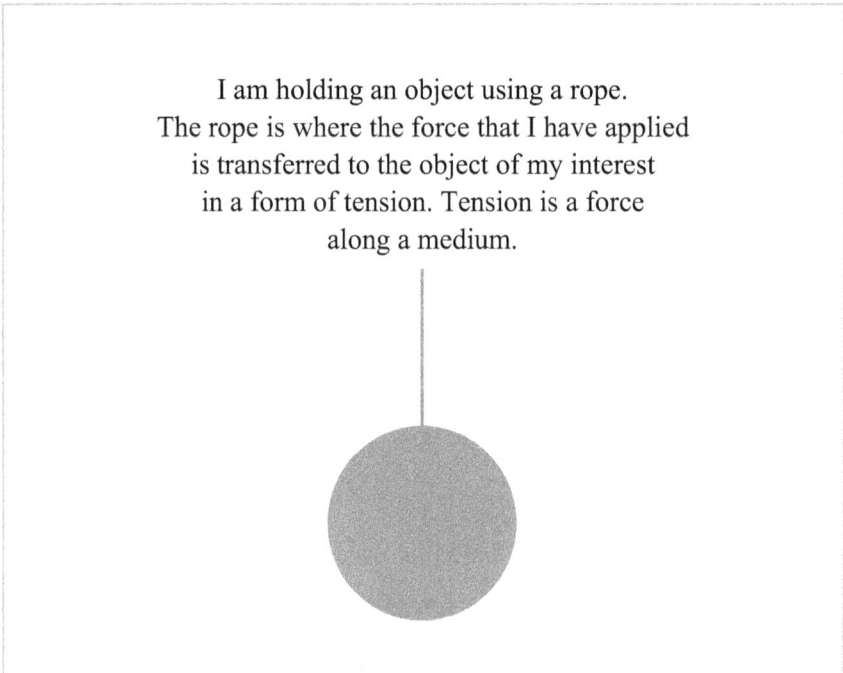

I am holding an object using a rope.
The rope is where the force that I have applied
is transferred to the object of my interest
in a form of tension. Tension is a force
along a medium.

Figure 23.1: This shows that an object could be held using a rope, where the force I am applying is transferred to the object along the medium, thus I do not need to directly hold the object. You are holding the rope, not the object directly. This is doable because of the tension that is transferred along the rope. Tension, by definition, is the force along a medium, material that has a certain degree of elasticity. Now, imagine that the object of our focus does not have a structure in space. For instance, it is a test particle. Do we need to worry about tension?

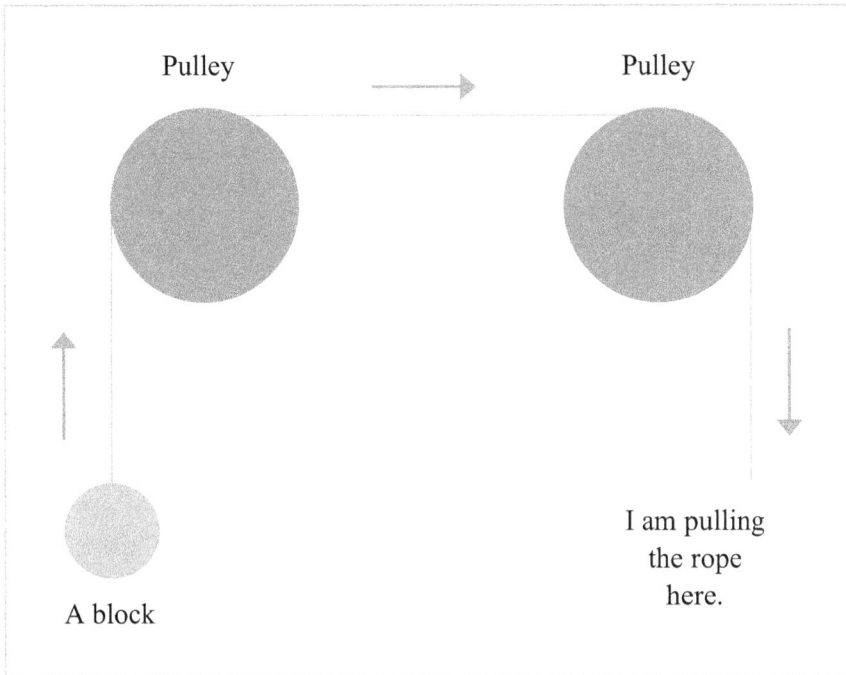

Pulley   Pulley

I am pulling
the rope
here.

A block

Figure 23.2: This illustrates another advantage of having and utilizing tension. It is not only allowing us to transfer force along a medium such as a rope but also to change the direction associated with the applied force. The rope is hanging from the pulley where the direction of the force changes, so the block is going to move up as I pull the rope down. You may change the direction of the force that you're applying, depending on how you set up the pulleys.

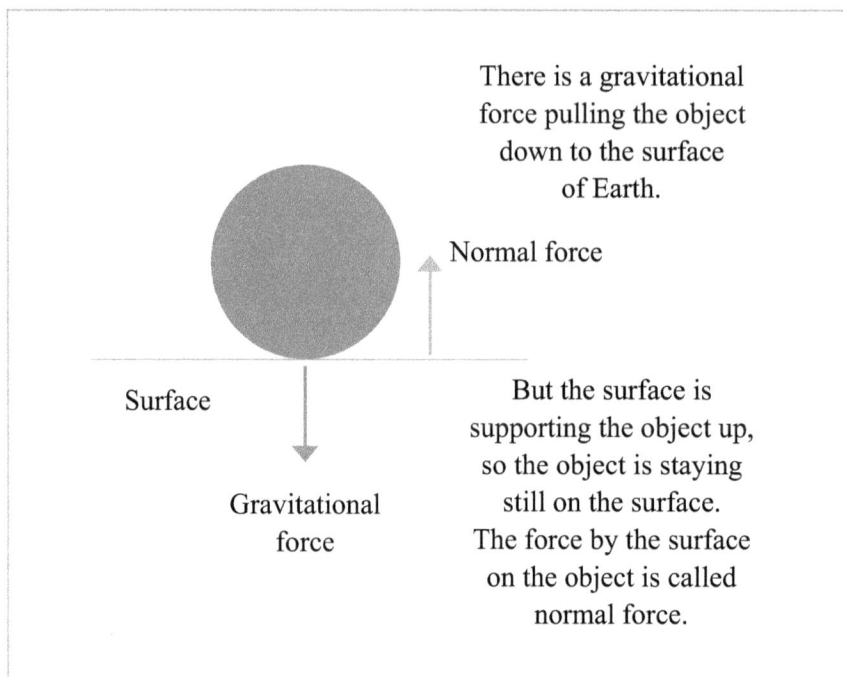

There is a gravitational force pulling the object down to the surface of Earth.

Normal force

Surface

Gravitational force

But the surface is supporting the object up, so the object is staying still on the surface. The force by the surface on the object is called normal force.

Figure 23.3: This illustrates what normal force is about in dynamics. If we have an action by the Earth, then there must be something that works in the opposite direction to support the object to stay on the surface. The two have to always be balanced out. The Earth is pulling the object down toward the very center of the Earth. If gravity were the only force on Earth, that is what would happen to the object. However, due to a force that keeps the object on the surface, the object stays on the ground. The force by the ground is the one. It has the same size of gravitational force but in the opposite direction. That is exactly what normal force is about. Remember that force always comes in a pair, the principle of action and reaction. There is no such thing as a single force being realized.

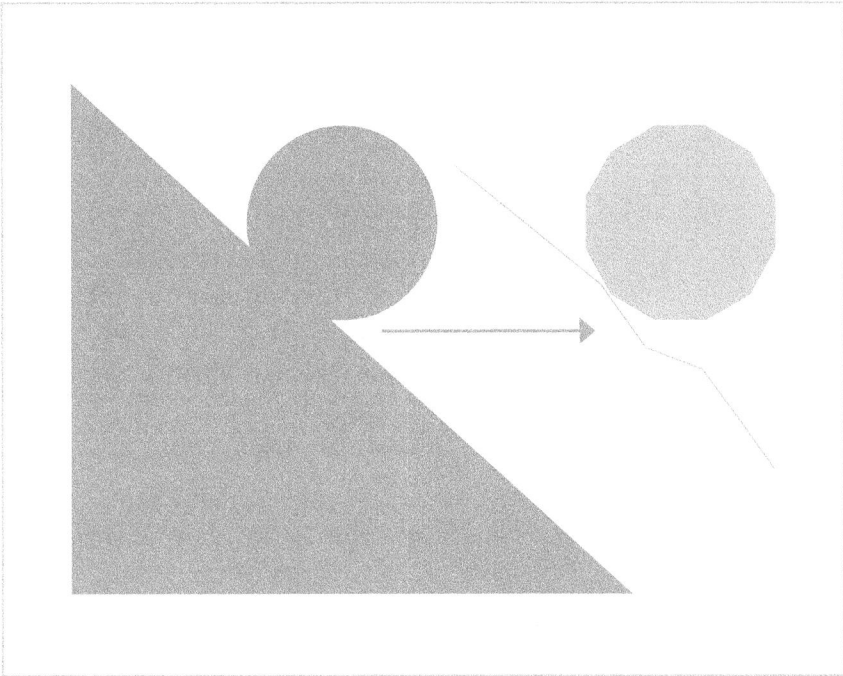

Figure 23.4: When we draw objects such as a sphere rolling down a plane with an angle, the left side of the figure is most likely what you would imagine. There is going to be an inclined plane with a sphere on top of it. However, the shapes associated with all the objects that we deal with in our daily lives are not that perfect. Nothing is perfect in physics. When you have a very close look at them and at the point where they appear to be in contact with another, the right side of the figure is what we likely end up with. There is no perfect sphere, but we have something that looks like a sphere with a finite number of sides. The inclined plane has all different kinds of wiggly shapes too in microscopic scale. Why is this important? This is what frictional force is about. The shape associated with any object occupying space is not perfect. Following the principle of uncertainty, nothing can be made perfectly; the position cannot be measured with perfect precision. Therefore, when an object is in motion and it is in contact with another object that is not in motion, the latter is going to resist the motion associated with the former. How large is the size of the force? That is the size of the friction force, and it happens to have two different types, depending on whether the object is in motion or not.

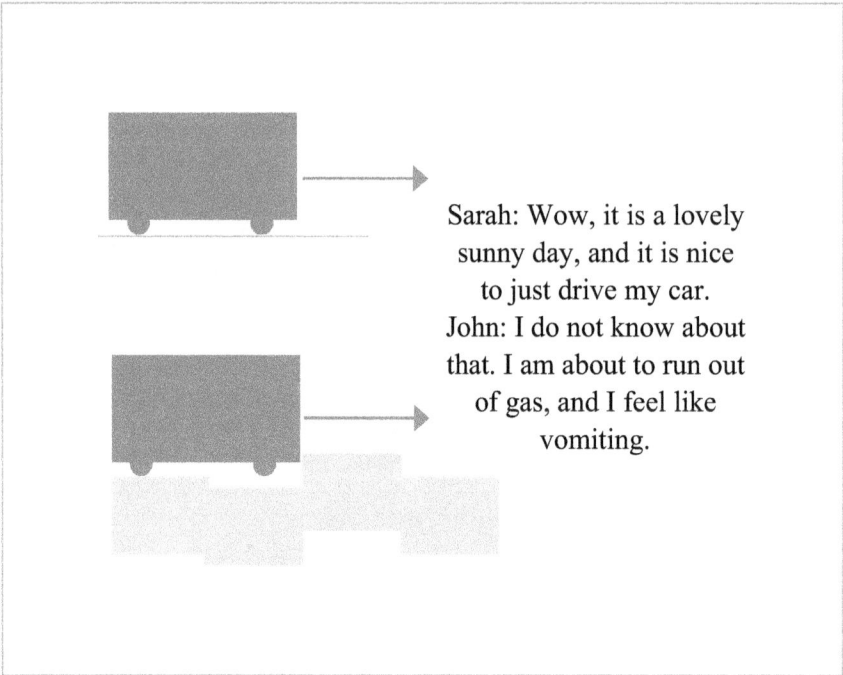

Figure 23.5: Imagine that Sarah is driving her car on a paved road, whereas John is driving his on another where we have all "wiggly zig-gly" shapes on the road, relatively speaking. Who needs to push the ped-als more often to maintain the original velocity that they began with?

Remember: when two objects with dimensions are in contact with each other at a microscopic level, we need to introduce normal force, friction force, and tension.

Problems:

Search for an article and find a case where the total force being applied to an object is greater than that of the force being applied by a person using a medium. For instance, it would be good if you could find a case of using many pulleys to lift an object up into the air. Find the case and relate that to the total work done on the object by the person lifting the object. Hint: you may go over the lessons we have in the latter part of this book.

The size of the frictional force happens to be independent of the area of an object or objects but depends on the type of materials. Think about how this is so. Briefly describe it in a paragraph.

Go over Figure 26.2 first. Calculate the force that needs to be applied by a person on the right side of the rope in order to keep the object up in the air. Compare the size of the applied force to the weight of the object.

Figure 23.5 illustrates two different roads. Imagine that the mass of the two cars is 100 kg including the drivers. The size of the frictional coefficient on the top road is 0.25, whereas the bottom one is 0.5. If the drivers could keep applying 1,000 Newtons of force all the time, calculate the velocity associated with the two cars after 10 seconds.

Day 24
Statics

"An object being at rest does not mean
that there is no force acting on it but that
the sum of all the forces, or the net force,
acting on the object happens to be 0, so
it appears to be at rest or in a constant
velocity. This is something different from
Newton's first law."

Question: you are staying at rest. Are there any forces being applied to
you or not?

You might have heard the phrase "delicate balance of life" in the
past. In classical mechanics, statics is all about analyzing such balances
when describing the state of an object or objects that appeared to be not
in motion. The "statics" represent a state where the object or objects stay
at rest, or they are in motion with a constant velocity with respect to some
reference. You might wonder why we study such a state. Is it relevant to
our understanding the dynamics in a system? Answer: Yes, it certainly is.
Understanding statics helps us to understand the "cause" part in dynamics
better. In short, statics in physics represent a state where an object has no
change in its velocity and no net force is acting on it, and we study how
the "delicate balance" is maintained.

Overall, there are two scenarios that we can consider: one where we
have a single object in a system and another where we have more than
one object in a system, and the object of our focus somehow maintains
a static state. If you remember Newton's first law, the former is not so
much of interest since it is just about an object in a static state from
which no other useful information can be extracted. Well, the only information I can tell you is that Newton's first law is all about that, so let us
not worry about the former too much. Also, you have already studied
Newton's first law.

That brings us down to the latter, a static state in dynamics, which
we can focus on and pay attention to. Again, it is not because the latter
has a different type of "apparent" state from the former but because we
can study and analyze why the object is in a static state by analyzing all

the force acting on it. Remember: it is mainly about analyzing the forces acting on it.

For instance, we can think of a simple example: there are three people standing up and aligned in a one-dimensional space. The two on the sides pull the person in the middle with the same amount of force. It is not too difficult to imagine that the person being pulled by the two people is not going to move. The person, if he happened to be at rest to begin with, is not going to be in motion. Why? Answer: The force exerted by the person on the left is canceled out by the force exerted by the person on the right side. It may sound simple, but it has an important implication as far as studying statics goes. Individual forces act on the person in the middle. It is not to say that there are "no" forces acting on the person. Forces are there. Remember that. But due to the fact that the two forces are acting in opposite directions with equal magnitude, no net force is going to act on the object, so the person in the middle is not going to accelerate. Figure 24.1 illustrates the point.

If you understand that simple case, that is exactly what statics in classical mechanics is all about. It is a "dynamical" static, not a "static" static. The latter, again, has something to do with Newton's first law.

> Statics in physics does not mean that there is no individual force acting on the object. There are individual forces acting on an object in dynamics, but it just happens to be that the total net force acting on the object is zero when we add them all vectorially.

Let us think about a more generalized case. An object is sitting in the middle, and there are four objects with equal mass placed the same distance from the object so that the total net force on the object happens to be zero Newtons. That means that in such a state, the object of our interest is not going to be in motion since the sum of the individual forces gets canceled out. When the forces are not balanced, the total net force on the object is not going to be completely canceled out, and the object will be in motion because of having a net acceleration. Yes, it is the amount of acceleration that we can calculate, and we can do what we did in kinematics utilizing the quantity.

There are many practice questions you can find on statics in dynamical states. You are encouraged to go over Figures 24.2 and 24.3 to understand the case of statics better, and you may find more practice questions later on. When you work on them, just remember that statics in classical mechanics is all about finding the quantities associated with objects that cause an object to be in a static state, or if it is not in a static state, then it is about analyzing and studying the kinematics associated with the object in an imbalanced state so that the state associated with the object in the future or in the past could be studied.

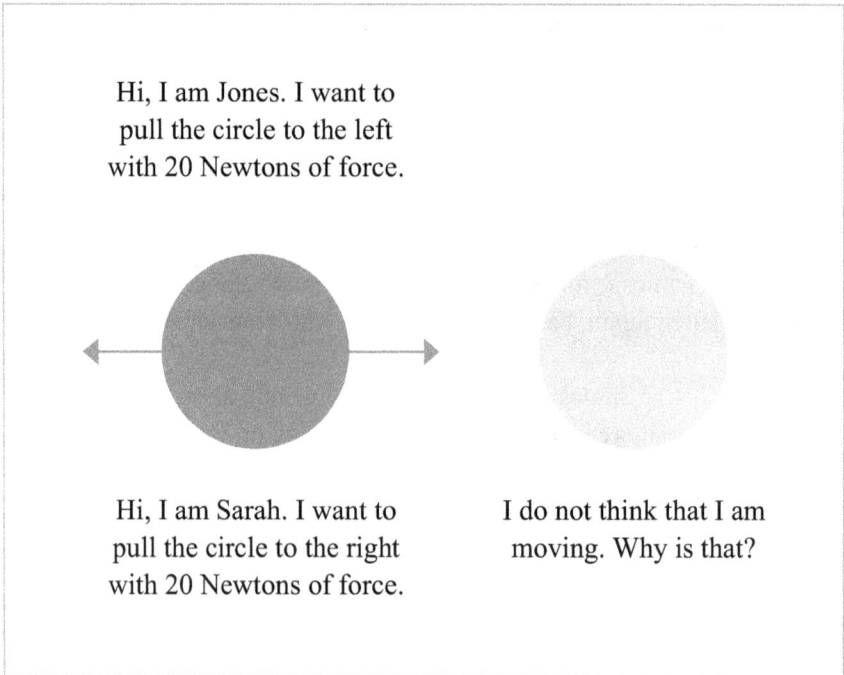

Hi, I am Jones. I want to pull the circle to the left with 20 Newtons of force.

Hi, I am Sarah. I want to pull the circle to the right with 20 Newtons of force.

I do not think that I am moving. Why is that?

Figure 24.1: This illustrates the "dynamical" statics. Someone tries to pull an object to the left with 20 Newtons of force, and another person does the same but to the right. If the two forces are the only two external forces acting on the object, then the object is not going to be in motion with respect to someone at rest. The "object" is going to think that it stays at rest, although we have a few external forces being applied. Point: having no acceleration does not mean that there is no force acting on an object. It is a dynamical state, but it just appears to be static.

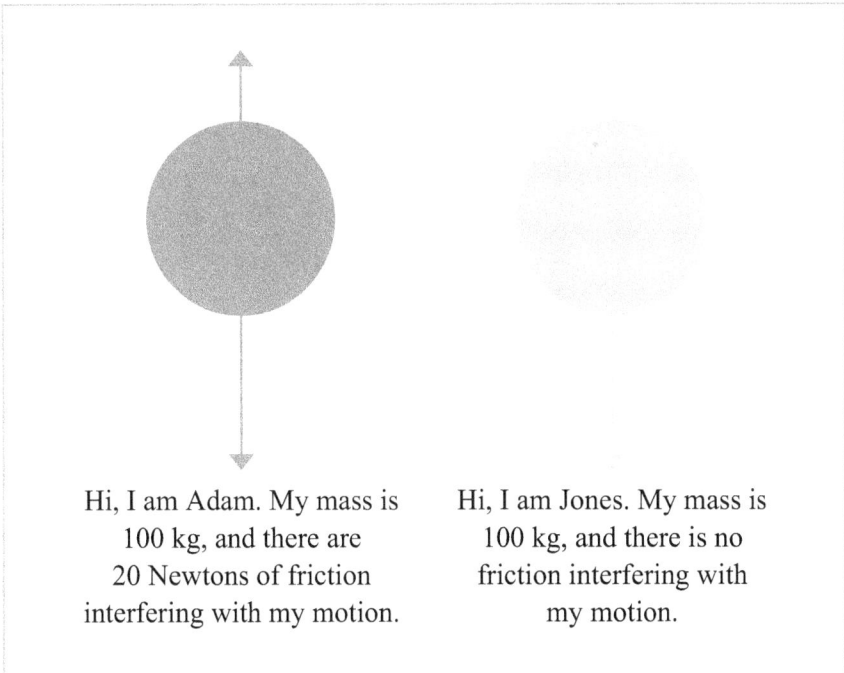

Hi, I am Adam. My mass is 100 kg, and there are 20 Newtons of friction interfering with my motion.

Hi, I am Jones. My mass is 100 kg, and there is no friction interfering with my motion.

Figure 24.2: Imagine that we have two 100 kg objects near the surface of the Earth, and they are falling to the ground. The object on the left-hand side has a frictional force acting up in the vertical direction, whereas the object on the right-hand side does not. Based on what we have learned as Newton's second law, can we calculate the acceleration associated with each case? Which one is going to have a larger acceleration?

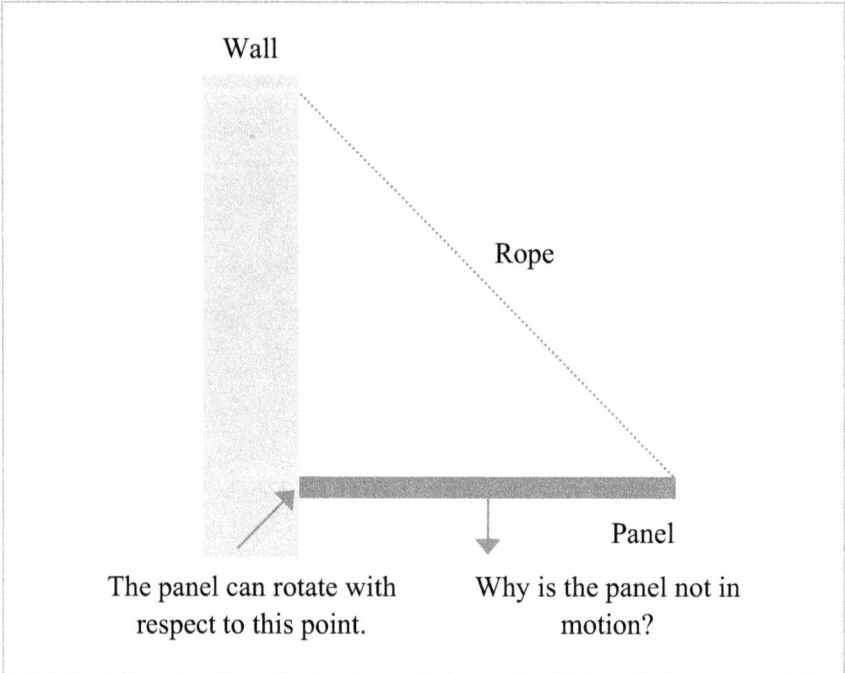

Figure 24.3: This illustrates another case of a static state. The left end of the panel is fixed to the wall but can rotate freely, and the right end is being held by a rope that is made of a flexible medium. The panel is not in motion. Why?

Problems:

Think about how we can stand still on the surface of Earth in terms of what we have studied as a static state and write a paragraph on it. You may need to study about the normal force and the fundamentals of reference frames in classical mechanics first.

Explain how an object that is static is related to Newton's second law. Briefly describe how the total net force on an object is related to a mechanical motion associated with a single object.

Can we ever realize a perfectly static state for an object in this universe? If so, describe how you are going to realize it. If not, briefly describe why you cannot do so.

Go over Figure 24.2 and calculate the size of the acceleration associated with the object in gray on the left and the object in light gray on the right. Note that Newton's second law is about the total size of the force that is acting on an object in a certain direction being the same as the size of the mass times the size of the acceleration associated with an object.

Go over Figure 24.3. Imagine that the mass of the panel is 10 kg, and we assume that the rope has a 30-degree angle with respect to the horizontal direction. Calculate the size of the tension on the rope that is needed to hold the panel so that it is at rest.

Day 25
Pressure and strain and stress

"Objects have dimensions, so the same
force can be delivered to different areas.
Objects have certain structures in space.
Because of this, we need to measure what
is called 'pressure.' Understanding strain
and stress is going to a bonus."

Imagine that you are going to hold a long piece of rod in your right hand and push your left hand hard using the rod. Then, imagine that you are going to do the same thing but using a bulky plastic box. In other words, the former is about poking your hand, and the latter is just pushing your hand with an object.

Now, here comes an interesting question: which case will cause you to have more pain on your palm, the piece of a rod or the plastic box? It may not take that much time to think of an answer. You will have more pain in the case of the rod, most likely. Would that be the case even if you applied the same force when using either object? Answer: Yes, it would be. Why? Because the size of the "pressure," which represents the size of the force as a function of the area to which the force is being applied, is going to be different. The former is almost the same as poking your palm, whereas the latter is merely pushing your hand with another object.

So, understanding and studying force in dynamics is not the end of the story. Why? Objects occupy space; they have dimensions, and thus something that has to do with the size associated with them needs to be addressed when studying classical mechanics. We try to be as specific as possible when analyzing motions or the causes of the motions. The same force can be realized but with a different size of pressure.

Physics is about understanding different states associated with an object or objects. We want to be as specific as possible. The same force can be realized on objects that are a different size, and thus we need to think about their having different sizes of pressure.

Mathematically speaking, pressure is a physical quantity that indicates the amount of force being applied to a unit area. So, going back to the case in the beginning of this lesson, when you push your hand with a piece of rod, which has a smaller area, your hand is going to feel more pressure compared to if you push with a plastic box, which has a larger area, relatively speaking. The same amount of force is delivered to your hand, but the sizes of the area associated with the objects are different and they play an important role. Pressure can be different when the amount of force is the same.

> Pressure is proportional to the applied
> force and inversely proportional to the
> size of the area on which the force is
> applied. The larger the size of the force,
> the larger the size of the pressure. The
> larger the size of the area, the smaller the
> size of the pressure.

So, like with our introducing acceleration, a change of velocity being a function of time, we just come up with another quantity, which happens to be as useful for dividing two useful physical quantities.

Figure 25.1 and Figure 25.2 illustrate the main point. Figure 25.1 is where a plastic block with a spherical shape pushes an object with a rect-angular shape. The block has a larger area, so it won't distort the shape of the rectangle much, even if you are pushing it very hard. With all proba-bility, it is hard to realize. Imagine that you have a phone and try to make a hole on a piece of paper. Is it going to be hard or not? On the other hand, what if you do the same thing but with a long piece of a solid wire? That is what is illustrated in Figure 25.2. When a rod hits the rectangle, it is going to go through the rectangular object. The same amount of force is applied in both cases, but one of them with a smaller area, which deliv-ers more pressure to the object. In short, pressure is proportional to the applied force and inversely proportional to the size of the area on which the force is applied. The larger the area, the smaller the size of the pres-sure when the size of the applied force happens to be the same. Again, having the same size of force is not the end of the story. For test parti-cles, it might be the end of the story since they do not have structures in the spatial dimension, but for all the objects that we deal with on a daily basis, they do have dimensions in space, and so we have to deal with the pressure to be more specific with our analyzing the state. By the way, the word "stress" is another name for pressure.

You might ask why pressure is such an important quantity in classical mechanics. Answer: It is because pressure has to do with the rate of change of the shape associated with an object as a function of its original size. Think about it: if pressure is what can be identified as the action, there must be a reaction. So, here comes another important physical quantity and its name is "strain." Again, it is to represent the reaction side of our applying the pressure.

Strain, by definition, is the change of length or size of an object with respect to its original length or size. For instance, when the length of a rod is 100 cm but the change in length was 1 cm because of pressures that are being applied to the object, then the strain associated with the rod in this case is going to be 0.01, which is the rate of change in the length with respect to the original size. In short, the larger the strain, the larger the rate of change of the size associated with an object is going to be. Figure 25.3 illustrates the point. It is not too hard to imagine that it would be easier to distort the shape of a rubber glove than of solid aluminum. Point: the degree of distortion is different depending on the medium where the force is applied.

So, we now have two quantities, stress and strain, or pressure and strain. The former indicates the force per area, and the latter represents the rate of change in shape with respect to their original shapes. In terms of cause and effect, stress is the cause and strain is the effect. Remember that.

Stress is the cause and strain is the effect. Stress needs to be applied for strain to be realized.

Taking all that into account, we can think about a slightly more complicated case. We can apply different amounts of force to objects made of different material. But you know what? Amazingly, there is a unique ratio between the size of the stress and the size of the strain per medium. What is this all about? Yes, here comes the famous "Young's modulus," or what is called the degree of elasticity, the rate of stress per strain associated with a material—something that can be tested and is uniquely associated with mediums. In classical mechanics, Young's modulus is an important quantity since it indicates the degree of what we understand as endurance associated with a medium because of our applying an external force. In practice, for instance, this is going to be important when people construct a large and tall building. Young's modulus associated with materials in the field of architecture should not be so large that it breaks. At the same time, it should not be too small, so the shape doesn't get too distorted when a certain size of external force is applied. Again, all these are something that we need to worry about when dealing with an object that occupies space, not for a test particle.

In short, stress is force per area, and strain is the percentage change in its shape; thus, both quantities gain their meaning when objects have dimensions. The ratio of the two is Young's modulus, and it is a quantity that is useful when material with a certain degree of elasticity needs to be utilized. Remember that we just want to be as specific as possible when analyzing mechanical motions.

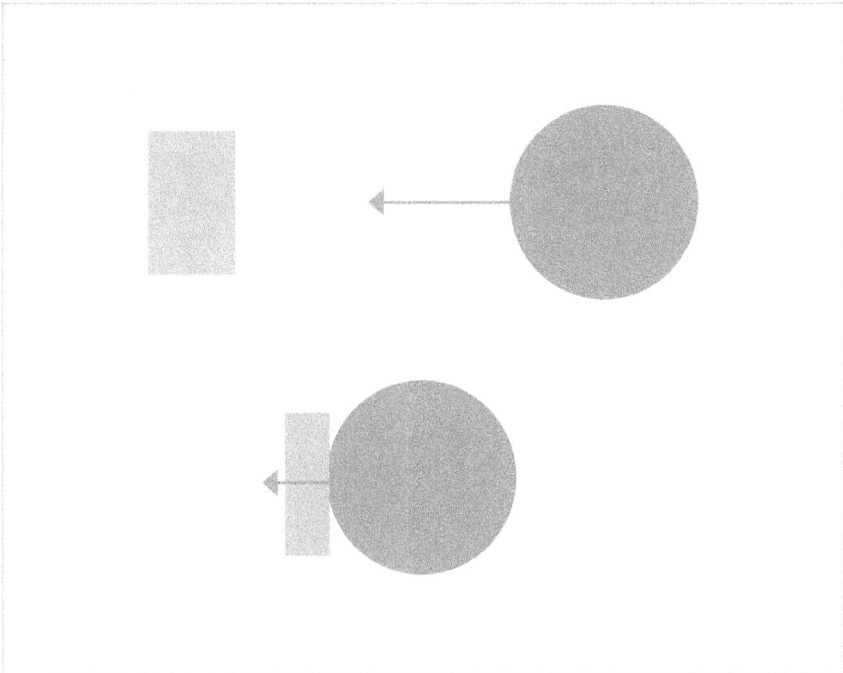

Figure 25.1: When a plastic block with a spherical shape hits a rectangular object, it is going to distort the shape of the rectangular shape object a bit.

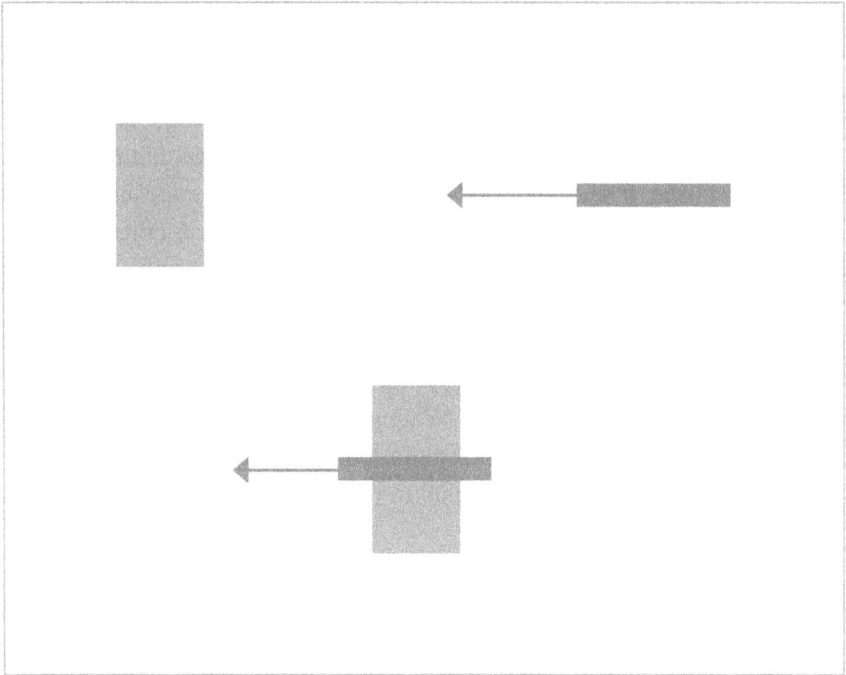

Figure 25.2: On the other hand, when a rod hits the rectangular object, the pressure that is delivered by the rod compared to the plastic block in the previous figure is much larger, so it not only distorts the shape, it even goes through the object. The same force is applied, but the mechanics is different, depending on how much pressure is delivered by a certain object.

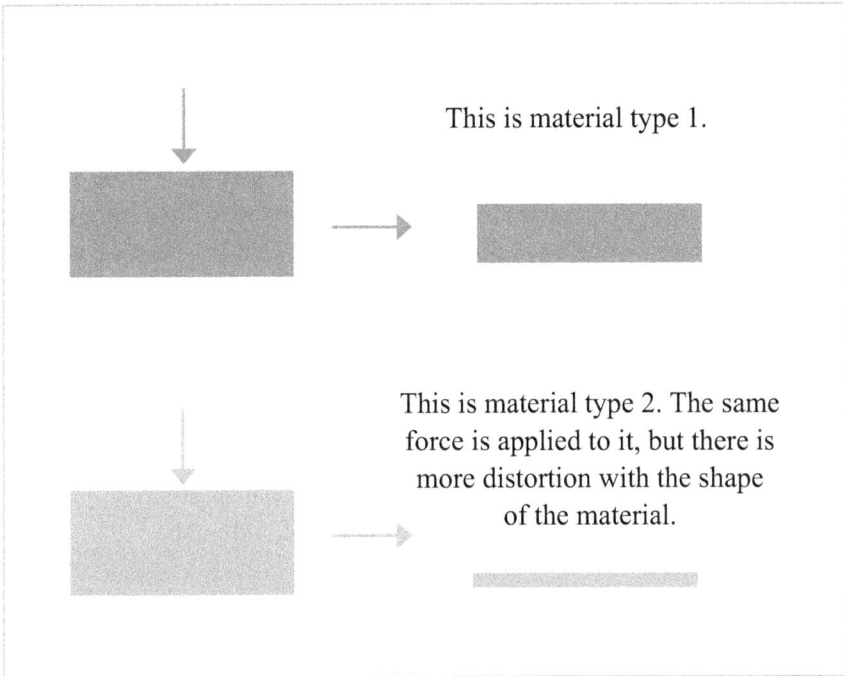

This is material type 1.

This is material type 2. The same force is applied to it, but there is more distortion with the shape of the material.

Figure 25.3: When the same amount of force is applied to two different materials with the same area, the change in height of the material is going to be different depending on the type of materials. Here, the strain associated with material 2 is larger than the strain associated with material 1, so the change in height for material 1 is greater. The pressure per strain, or the stress per strain, or what is called Young's modulus, is going to be a unique quantity for different materials.

Remember: in physics, we want to differentiate motions by physical quantities. If two motions have different values in any combination of physical quantities that can be derived from the fundamental quantities, including displacement, time, and mass, we do want to treat the two motions as different motions. The same goes here too. When a different amount of force is applied to a different area with a different type of medium, we want to differentiate them. That is where we need to introduce pressure, stress, and strain—thus, Young's modulus.

Problems:

Can Young's modulus be defined in one-dimensional space? If so, describe how it could be. If not, describe how it could not be.

Pressure gains its meaning and importance only if the object on which the pressure is applied has dimensions. In other words, we cannot think of the quantity of pressure being used when the object happens to be a test particle, an entity that does not occupy any dimension. Think about this in terms of force and area.

You may notice that the size of the arrows in the bottom of Figure 25.2 and Figure 25.3 are smaller than the ones on the top. Think about why it needs to be so and what type of force was covered in previous lessons to take them into account.

Day 26
Energy

"Change in energy is defined as work
done to or by a system. When an external
force is applied to an object or objects,
there is going to be a change in energy.
Otherwise, the energy stays the same."

It is hard to describe what energy is about, but we can think of it as the following: energy within a system does not go anywhere. No matter what that is, that does not change as a function of time and there it gains its importance.

Kinematics is all about understanding a mechanical motion associated with an object or objects in a system. In essence, we want to be as specific as possible when describing the mechanical motion, thus we introduced some useful quantities such as velocity and acceleration. After that, we studied how they relate to each other. Problem: The physical quantity in kinematics does not get conserved. It changes. Think about velocity and acceleration for a moment. When something is thrown up in the air, does the size of the velocity and acceleration change or not?

So, what do we do? What do we do to study something that may not change over time? Where do we study something like that?

Answer: We study dynamics, if not kinematics. However, understanding dynamics is somewhat different from understanding kinematics. For instance, we study the cause by which a motion is being realized in dynamics. Particularly, we introduced a quantity called "mass" to differentiate an object from others, and that led us to understand dynamics better by studying what we understand as force.

Is that the end of the story? No, it is not. We need to study at least one or maybe two more. They are energy and momentum in dynamics. We are going to study energy first. It is going to be more interesting as we get into the topic further, but it could be a harder topic to deal with, so let us focus hard on this topic.

Let us go back to Figure 26.1 briefly. We had studied the figure when going over the tension in dynamics. Figure 26.2 shows something interesting. In the figure, we have a total of three pulleys, where two of them are fixed on the top portion of the wall, but in the middle, where the 100 kg object is directly attached, it is allowed to be in motion freely. The person standing on the ground in the figure pulls the rope down. That is how things were set. Question: how much force does the person need to exert when pulling the rope? Instead of pulling the rope with 1,000 Newtons, which is the weight associated with the object, the person needs to exert only half of it because of the way the pulley is set up. There are going to be two pieces of the rope that are going to support the pulley in the middle, so the sum of tension on the pieces needs to balance the weight of the object. For that reason, the tension needs to be only half of the weight, thus the force that the person needs to exert would also be half of the weight. This was what the tension lesson was about.

Question: how come that happens? Is it not weird? Nature happens to be fair. So, how come the person exerted only half of the weight of the object and was allowed to pull it up? Is it not unfair in terms of the physics that runs behind it? Again, we were told that physics is fair when studying the principle of action and reaction, and we all know that. Is it magic? Answer: No, it is not. However, again, remember that physics is always fair. Thus, if we were to apply a lesser amount of force to the rope and were able to pull the object up in the air, then our applying less force would always need to be "compensated" for somewhere else.

Physics is always fair. That means that if you apply less force, then you applying less force needs to be compensated for somewhere else.

If something is measured or reported less than what is expected, then something else needs to become larger in order to keep the size of the quantities within the system as a whole the same as before.

Nature abhors change.

Nature wants our system to stay the same as before. If we apply a force that is less than the weight of an object, this application of less force needs to be compensated for someplace else, or by some physical quantity. Now, can you think of something? Think hard on this one.

Here is a small hint that you can consider: You may need to lift "both pieces" of the rope that is attached to the pully in order to lift the object up using the pully system. What does that mean? Answer: You need to pull the rope more than you used to when you are directly pulling the object up. In other words, you need to displace the rope more than the displacement associated with the box. Do you see the point here?

We need to pull the rope down more.
In other words, we need to displace the
rope more than the object that is going to
get displaced.

Yes, that is correct. You apply less force and that needs to be compensated for by pulling the rope more. In this case, we need to pull the rope farther by twice as much as the object that is going to be displaced. Keep in mind that there are effectively two pieces of rope that are pulling the pulley in the middle. For instance, if the object is to be lifted 1 cm up in the air, we need to pull the rope down by 2 cm; you need to pull the rope down two times longer than the size of the displacement of the object hanging in the middle.

Point: this is all about our understanding "conservation" of energy, or mechanical energy, in physics. Also, this is the essence of what energy is about in physics—not defining them using mathematical terms but illustrating them with figures.

As mentioned in previous lessons, force is not conserved within a system. You just apply less force. However, the size of the total work done by the force is going to stay the same before and after. The work done to a system in this case can be defined as the force times the displacement associated with the object. More precisely, it is a "dot product" of the force and the displacement, if you are interested. The work done on the object needs to be the same as the work that is applied to the object. In other words, the force times the displacement in terms of the motion associated with the object and that applied by the person is going to be the same. The force is different, but the force times displacement is the same. They need to be the same for the total energy within the system to stay the same. Question: how much work is done on the object by the rope? Answer: 1,000 Newtons multiplied by 1 m; thus, it is going to be 1,000 joules, the unit for energy. Question: how much work is done by the person pulling the rope down? It is 500 Newtons multiplied by 2 m, so the size of the work done is the same as 1,000 joules. Point: the total size of the work done applied to realize a certain state in an isolated system is going to be the same. The size of the force is not going to get conserved, but that times displacement, or the dot product of the two, stays the same before and after some interaction takes place.

For now, let us just understand what energy is about that way. Let us not get into too much mathematics. It is something in the units of the dot product of force and displacement; it is not just by the force, not just by the displacement, but by the combination of the two. Figure 26.3 is shown to help you understand the illustrated story better.

> Total energy in an isolated system is
> going to stay the same unless there is an
> outside force acting on the system.

Is that it? There is a bit more to be covered. Believe it or not, there are different types of energy that we need to consider, one that has to do with motion and another that has to do with the state. We are going to go over them in the next lessons. For now, let us just understand that energy is something that is conserved, in total. Again, energy does not go anywhere, and that is where it gains its importance. That being said, the total energy in our universe is going to stay the same.

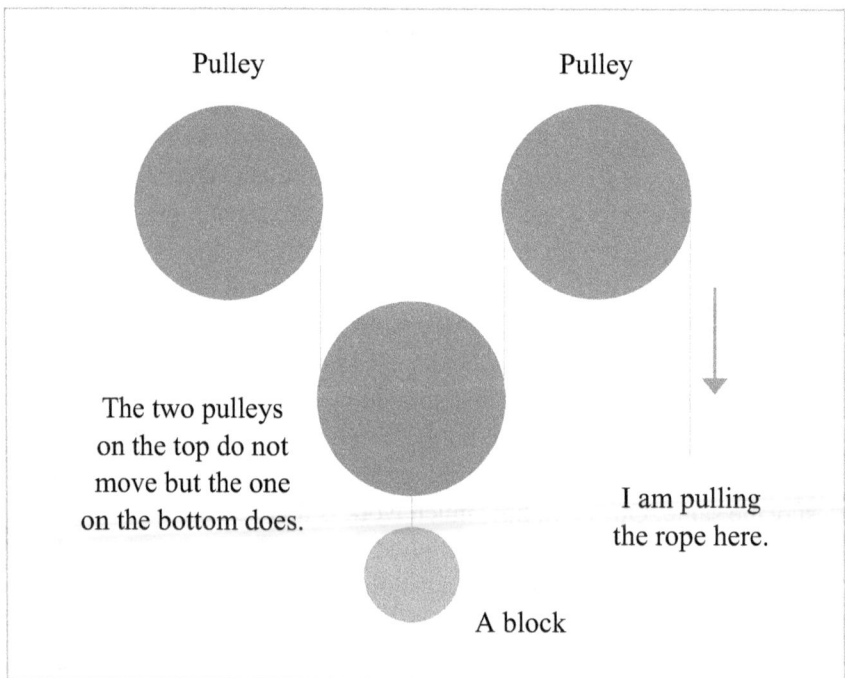

Pulley   Pulley

The two pulleys
on the top do not
move but the one
on the bottom does.

I am pulling
the rope here.

A block

Figure 26.1: This is to illustrate the advantages of using pulleys and ropes to lift an object up or down. Assuming that the weight of the block is about 1,000 Newtons, what do you think is the magnitude of the force that I need to apply to the rope on the right side in order to keep the block up in the air? Is it going to be 1,000 Newtons or is it smaller than that?

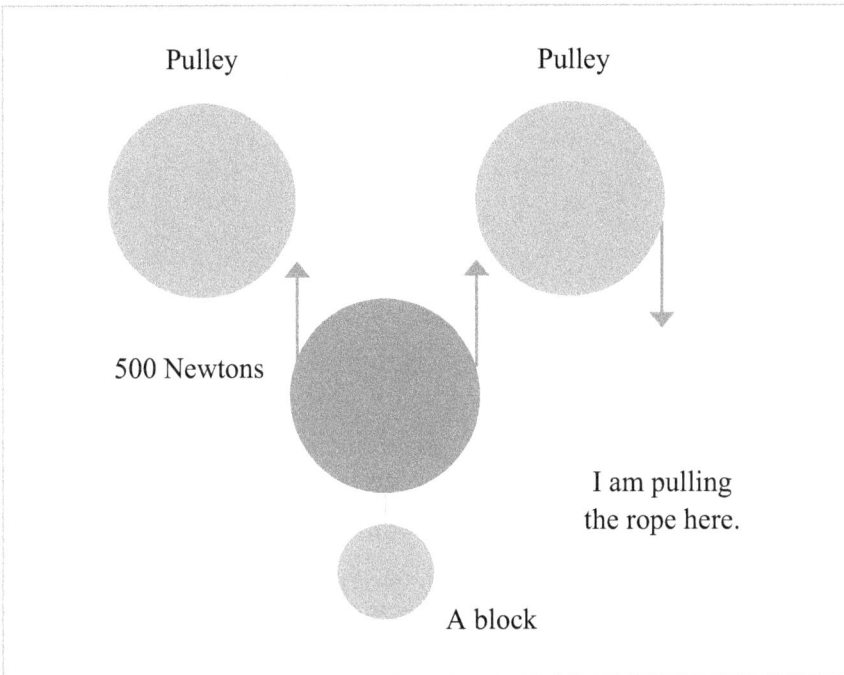

Figure 26.2: The tension is applied along the rope, so the weight is going to be equally divided into the two pieces of rope as shown in the figure; the gray part of the rope is going to have 500 Newtons of force applied to it, instead of 1,000 Newtons, since two pieces of rope are pulling the object upward. So, the bottom line is that I need to apply only 500 Newtons of force in order to keep the object up in the air. But nature is fair, so something needs to compensate for that. It is important to understand the point.

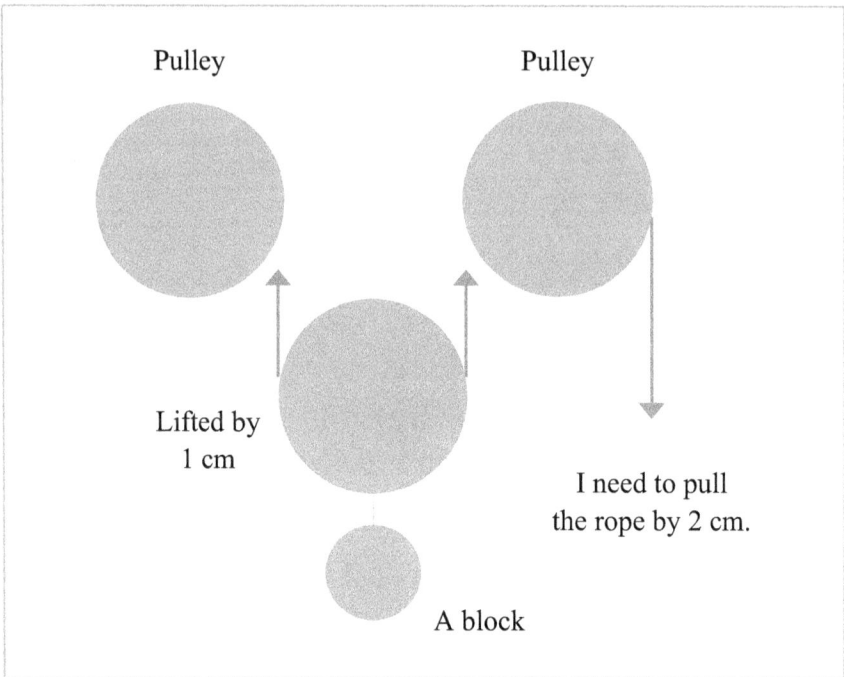

Pulley   Pulley

Lifted by
1 cm

I need to pull
the rope by 2 cm.

A block

Figure 26.3: This one is important to understand since this illustrates the connection of your understanding of kinematics to your understanding of dynamics. If only 500 Newtons of force is needed, since two pieces of rope are pulling the object up, then you need to lift the rope by 2 cm on your own to lift the object up by only 1 cm. Why? It is because both of the gray ropes need to be lifted by 1 cm, so you, at the end of the rope, need to pull the rope down by 2 cm. If less force is applied, the "less" force needs to be compensated for from somewhere. Remember, physics is fair, always. If something gets reduced, there needs to be compensation someplace else.

Remember: energy is a physical quantity in the units of the dot product of force and displacement. It is not just about force. It is not about displacement only, either, but the combination of the two. Keep in mind that the energy we deal with in classical mechanics is in the units of force multiplied by displacement, no matter how different the types of energy look to each other.

Problems:

We have two different types of mechanical energy in classical mechanics: one that has to do with kinematics, and another that has to do with state. Find them and define them in terms of other kinematical and dynamical quantities in classical mechanics.

Energy is known to be a "path independent" quantity, but force is not. Think about why this is so and briefly describe your thoughts.

Day 27
Kinetic and potential energy

"Kinetic energy has something to do with
the velocity, whereas potential energy is
something to be realized later."

We have two types of mechanical energy.

Let us go back to the case where we have a pen in Charlotte and a piece of paper in Columbia. We just assumed that they happened to be there to begin with. In other words, we did not think about how it happened to be. In any case, once the two objects are in the two cities, they will start moving; the pen is going toward the city of Columbia and the paper is going toward the city of Charlotte. The two objects pull each other gravitationally, and that is how the motion starts.

Let us go back to the initial state for a moment, where both the pen and the paper had no velocity; they were not in mechanical motion. Question: what does that mean? It means that they did not have "energy" associated with their motion. Does that mean that the objects as a whole in the system had no energy associated with them at all?

Answer: It depends. It really depends on what types of "energy" we are discussing. When we have a test particle being placed within a system and it is not in motion, then there is no mechanical energy associated with the object in a classical regime. Think about Newton's first law, the principle of inertia, and that is exactly what that is about. Figure 27.1 illustrates the point again for your review.

However, in dynamics, the story goes somewhat differently. What if we introduce one more object in the system? Point: all we did to transition from kinematics to dynamics was to bring one more object into the system where the original "single object" was placed. As shown in Figure 27.2, we have two or more objects in the system. Then, what is going to happen? Yes, they are going to interact with each other gravitationally; they are going to pull each other.

In the very beginning with the two objects, they were at rest. But you know what? Surprisingly, there still is a type of mechanical energy that we need to consider. Can you think of one? Yes, that is what potential energy is about. It is a type of energy that can be realized in other forms of energy "later" but is currently represented by the distance between the two objects. So does the word "potential." It is not realized in terms of a mechanical motion.

> Potential energy is a type of mechanical energy associated with the distance between two or more objects. It is not something that is realized in terms of a motion. When two objects are at a distance, the energy that can be realized later in terms of the velocity, or the motion associated with an object, can be quantified by the distance between them.

Potential energy is not something realized or manifested in terms of motion, so potential energy can be thought of as a type of energy that is being "latent" as a function of the distance to other objects in the system. Again, you have two objects at a distance, and they are not in motion at the moment. However, due to their interacting gravitationally, we can expect that they are going to be in motion in the future, and we can quantify the degree of their being in motion using the distance between them as of now. That is exactly what potential energy is about. It is a function of the distance, and that is how we can quantify the size of the mechanical energy to be realized later.

On the same line of thought, kinetic energy can be thought of as a type of energy that is "already" realized as a function of kinematical quantities. In other words, kinetic energy can be quantified as a function of the object's velocity. Of course, we need to introduce mass as a main factor in their expressions. For their mathematical definition, you may want to refer to other sources of literature.

You may think of a reason to bring the definition of energy from a different perspective. Say you have two objects, and they are at a different distance with respect to the Earth in the vertical direction. Question: how do we distinguish the state of the two objects? The potential energy of the gray object with respect to the Earth is larger than that of the light gray object since the gray object is farther from the surface. The longer the distance, the larger the potential energy. Potential energy is where the two objects can be differentiated. Figure 27.3 illustrates the point.

Coming back to the case where we have a pen in Charlotte and a paper in Columbia, they have mass, so there is gravitational interaction between them, and their interaction is going to cause the objects to move toward each other, so the objects are going to gain velocity. When the motion of objects begins, the objects gain kinetic energy. Here comes an interesting question: from where did they gain kinetic energy? They gained it from the potential energy that they started with. As the potential energy gets smaller because of the distance getting smaller, the kinetic energy gets larger. Does that ring a bell? What does that mean?

Yes, that is what "conservation" of energy is all about in classical mechanics. The total mechanical energy within the system, the sum of kinetic and potential energy, is going to stay the same before and after. For instance, if an object is 100 cm away from the surface of the Earth, you can calculate the sum of the two energies. Then you do the same once the object is in contact with the surface. The sum in the former and the latter should be the same. Why? Energy does not go anywhere. The total is going to stay the same.

In short, just keep in mind that mechanical energy is in the unit of force times by displacement, and the sum of energy to be realized and that is already being realized is going to be the same within an isolated system.

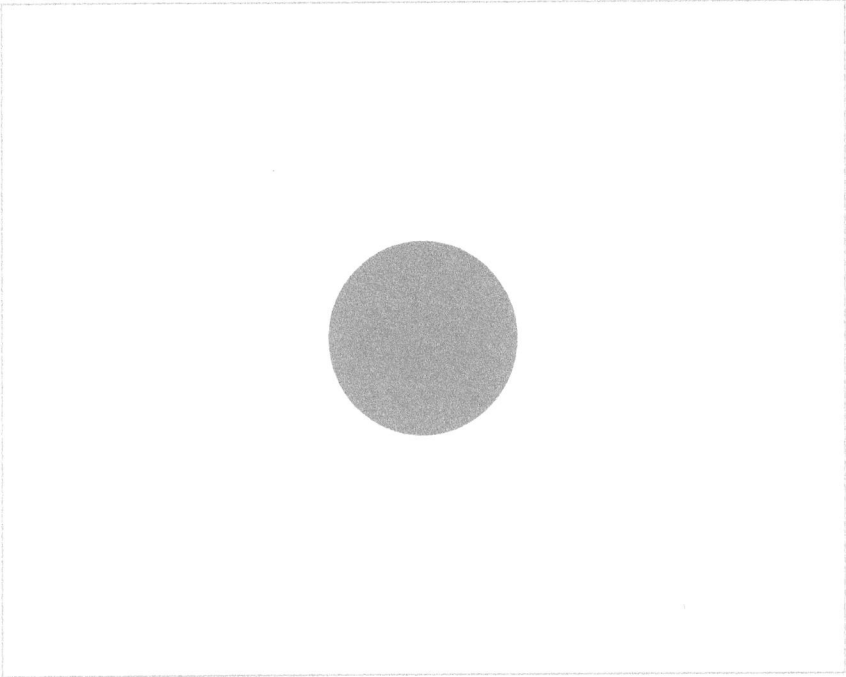

Figure 27.1: When we have a single object that is not in motion within a system, there is no mechanical energy, assuming that there is no internal energy that we need to consider. Why? Energy is a dynamical quantity that needs to be defined with respect to another object; if we have a single object, and that object only, then we cannot define "energy" in physics. However, if the object does occupy space, there is something that we need to consider. What is that about?

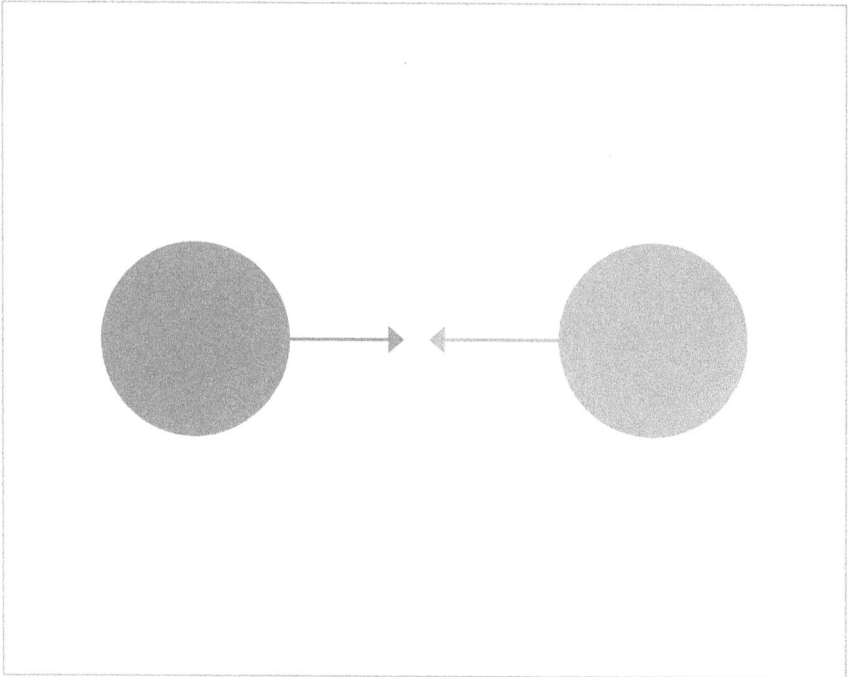

Figure 27.2: But as soon as we introduce one more object into the system, the two are going to interact with each other gravitationally. The moment the object in light gray is placed in the system, even for the moment that both objects are not in motion, there is a type of mechanical energy present. That is what potential energy is about; it is energy that is to be realized in other forms later.

There are two objects. The one
on the left is 100 cm away from
the surface of the Earth, and the
one on the right is 50 cm away,
and they are test particles.
How do we distinguish the
status of the two?

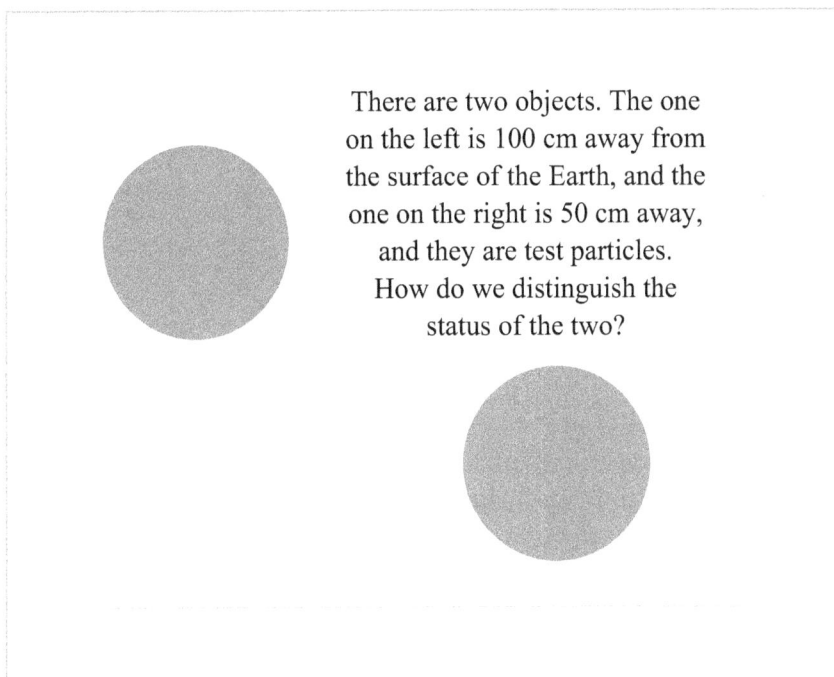

Figure 27.3: This is similar to what we had in a previous figure but with
the same mass and different positions with respect to a reference object.
One of them is 100 cm away from the surface of the Earth and the
other one is 50 cm away. We need to introduce a quantity called potential energy to distinguish the two objects in terms of their kinematics
and dynamics.

Remember: kinetic energy has something to do with the motion that is already realized in physics space. Potential energy can be realized later, but it is not manifested in physical space yet. If you add them up, you are going to end up with the total size of the mechanical energy, and that stays the same unless there is an external entity bothering the system.

Problems:

You have a single object within an isolated system, but the object occupies a space. Does the system have some energy? If so, describe what type of energy is associated with the system. If not, why not?

In practice, the total mechanical energy, the sum of potential and kinetic energy, in most isolated systems can be changed as a function of time. However, in this lesson, we mentioned that the sum of the total mechanical energy does not change before or after some interaction takes place.

Think about why this is so and describe your reasoning in a paragraph.

Find out the definition of potential and kinetic energy associated with a small object that is due to the presence of the Earth. Imagine that the object is about 100 cm away to begin with, not being in a mechanical motion, and then falling freely to the ground. Calculate the size of the linear velocity associated with the object just before the object will hit the ground. Hint: the size of the mass is not given. Why is that?

Day 28
Momentum

"Momentum stays the same within an
isolated system unless there is an external
entity or entities acting on the system.
A change in momentum will be delivered
to the other objects as force. Momentum will always be conserved, whereas
the total mechanical energy may not be
conserved always."

We have studied force and energy, so we have one more to go. What
is the last one that we are left with? Yes, momentum is the one. By
definition, momentum is mass times the velocity associated with it.
Mathematically speaking, we can write the size of the momentum as:

$$Momentum = Mass \times Velocity$$

We know that velocity is a vector and mass is a scalar, so momentum
is going to be a vector. It carries information regarding both magnitude
and the direction associated with the velocity but it is scaled by its mass.
There is one more thing to remember: what we study here as momentum
is linear momentum, something that we need to play with when describing a linear motion. We need to introduce another kind of momentum
when describing a rotational motion. We are going to study that later.

Is that all we need to know and study about momentum? Well, there
is more to it, and here we go. Total momentum stays the same within a
system unless an external force is acting on it.

Total momentum stays the same within a system.

In other words, total momentum within an isolated system does not
change as a function of time unless a force from an external entity or
entities makes some impact on those objects within the system. Does that
sound familiar to you? Hint: this is something that you studied early on.
The size of the velocity associated with an object does not change unless

there is an entity having an impact on it. It is basically the same as that, and that is because velocity is a component in the linear momentum.

Well, you may ask the following: What about the case where we have more than one object in a system, and they interact with respect to each other? In other words, what happens in a dynamic state? Answer: It does not matter what happens within an isolated system internally. As long as it happens within a system, the size of the total momentum associated within the system is going to stay the same. What does that mean? That is exactly what conservation of momentum is. For instance, if we have two objects in a system, then whatever changes that occur to an object must be delivered to the other one; thus, the total size of momentum stays the same as before. You can think of it as the balloon effect: one gets larger and the other one gets smaller, just like a balloon.

Also, there is one more important point to understand: force, a quantity to which Newton's laws can be directly applied, is something that we can quantify as a change in momentum as a function of time. You will see the point clearly in a mathematical expression. For instance, if you divide the expression above by time on both sides, we basically end up with what we have studied as Newton's second law. Keep in mind that force is not conserved in a system, but momentum always is.

> The total momentum within an isolated
> system stays the same unless there is
> an outside force acting on the system.
> Remember, that change in momentum
> is what is going to be delivered to other
> objects as force.

Let us go over a practical example before we move on. Imagine that you have 100 dollars as your salary to begin with and you spent 20 dollars of it to buy a notebook. In the end, you are left with 80 dollars in total. When you know the amount that you started with and the amount that you spent, then you can calculate the amount that you have left. Momentum is where this can be directly applied. Just like the amount of money does not go anywhere, momentum does not go anywhere. It stays the same within a system before and after. Does that ring a bell? Yes, the story is related to what we studied as Newton's first law. In other words, what we studied back in Newton's first law and second law is a part of what we are studying here in the linear momentum. It is just that

force is where Newton's laws can be applied, but they are related to
the momentum.

Now, let us think a little bit more about the conservation from a
microscopic point of view. Imagine that we have two test particles, and
they collide with each other. Particle 1 collides with particle 2, and the
size of momentum for particle 1 gets reduced as particle 2 gains some.
What does this mean? It means that momentum is going to be "transferred" from one particle to another, as velocity associated with particle 1 and particle 2 changes. Assume that the mass of particle 1 is 10 kg
and the initial velocity before the collision is 10 miles per hour and is
reduced to 5 miles per hour after the collision. Particle 1 loses some of its
momentum, and it is transferred to particle 2 within the system. Have a
look at Figure 28.1.

Question: how does that happen? How does the momentum of
particle 1 change after colliding with particle 2? Yes, particle 1 gets
"decelerated," so the velocity is reduced when it collides with particle 2.
The change in momentum associated with particle 1 with respect to the
time during which the collision takes place can be thought of as force.
Yes, this is a critical point to understand: the change in momentum is
going to be realized by force being applied to an object. Force is something that is needed in order to realize change in momentum.

Coming back to the case, the same degree of force is applied to particle 2, following what? Yes, it is the principle of action and reaction,
Newton's third law, if you remember. That is where particle 2 gains
more velocity and thus has a larger momentum than before. Again, force
needs to be a player in this scenario, and we can calculate the size of the
changes associated with the momentum. In other words, force is something that we need, and it helps us to quantify the change of momentum
in a system.

Point: when you read this lesson thoroughly, you may notice that
we have mentioned all three of Newton's laws when describing the
conservation of momentum. Yes, in order to realize the conservation
law of momentum in a dynamic state, we do need to utilize all three of
Newton's laws.

The conservation of momentum can be realized by using all three of
Newton's laws.

In summary, momentum is a physical quantity that stays the same within an isolated system unless an outside entity impacts the objects within the system. Do note that the quantity "momentum" is related to force in that the size of the change in momentum is realized as force transferring the momentum from one to another. Force is a mediator in this case. Keep in mind that momentum is only a quantity that is conserved no matter what type of mechanical collision we deal with. Force is not conserved. That is one of the disadvantages. Some type of collision is where total "mechanical" energy does not get conserved. The "total" energy within a system gets conserved though.

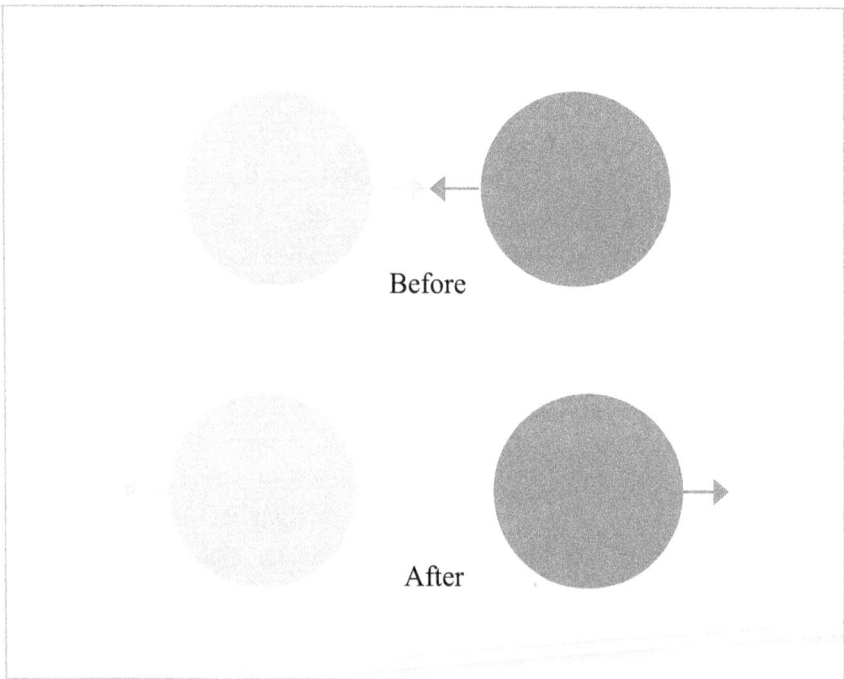

Figure 28.1: This illustrates an example of the elastic collision. We have two objects moving toward each other with a certain velocity. After they collide with each other, they move in the opposite direction. That one on the top left was moving to the right, but it is going to move to the left after the collision. On the same token, the one on the top right was moving to left, but it is going to move to the right after the collision.

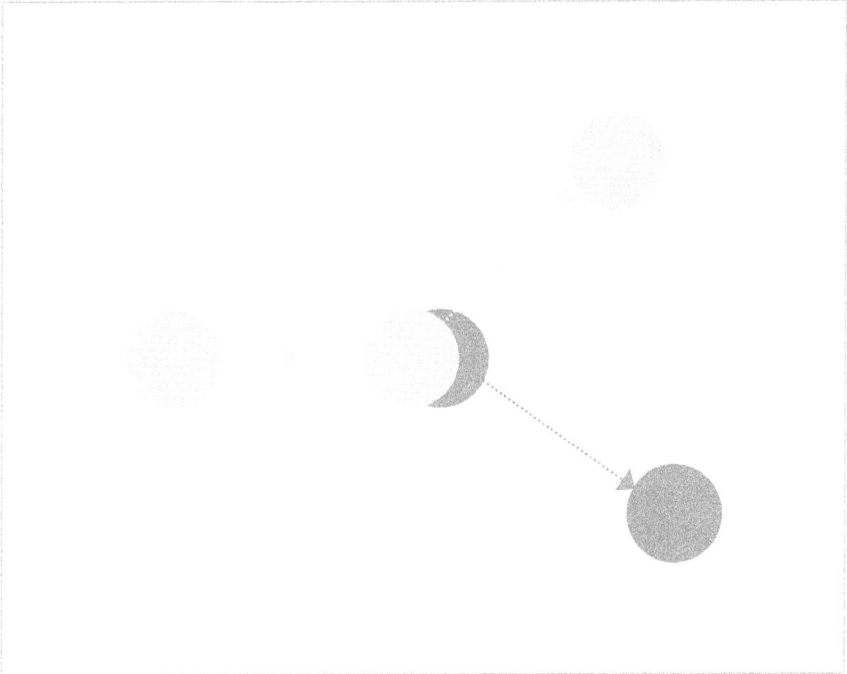

Figure 28.2: This illustrates an elastic collision between two objects. The sphere in gray is twice as heavy as the light gray one. The gray sphere was at rest, and the light gray one was moving 5 miles per hour. After the collision, the light gray sphere moves at a velocity of 2 miles per hour at 45 degrees with respect to the horizontal direction. Can we calculate the velocity and the angle associated with the gray sphere after the collision? If not, why not?

Remember: try not to think too hard about the mathematical definition of momentum, but think about why it gains its importance instead. Momentum is a physical quantity that stays the same whether it's an elastic or inelastic collision between two objects that takes place, so it is a critical quantity to be utilized when calculating the size of physical quantities when some quantities in the initial state are known or calculated. We can do some predictions based on that. Why not use force? Well, force is not conserved in a dynamical state.

Problems:

Can we ever create an isolated system where we can test conservation of momentum perfectly? If not, why not?

Go over the equations in this lesson, find out a definition for momentum, and derive Newton's second law from the definition taking some assumptions covered in this lesson. Remember that the change in momentum as a function of time is just what we have studied before when going over Newton's laws. Think about why this is the case and describe its implication.

Force is where Newton's laws can be directly applied, and momentum is where they can be indirectly applied. Briefly describe why this is so.

Go over Figure 28.2. The sphere in gray is twice as heavy as the light gray one. The gray sphere was at rest, and the light gray one was moving 5 miles per hour. After the collision, the light gray sphere moves at a velocity of 2 miles per hour at 45 degrees with respect to the horizontal direction. Can we calculate the velocity and the angle associated with the gray sphere after the collision? If not, why not?

Day 29
Elastic and inelastic collisions

"The total mechanical energy in a system is not going to change in an elastic
collision but in an inelastic collision. But
momentum gets conserved no matter
what type of collision we deal with."

Understanding momentum is so important in classical mechanics, so let
us go over them further. If you are interested in studying physics more
and more, you will study "generalized momentum" later on in a generalized coordinate system. For now, let us focus on linear momentum.

Imagine that someone throws a chunk of mud at my face because
he or she just wants to see how the shape of the mud is going to change
as it hits my face. Then what happens? Well, the shape of the mud does
change. It is not too difficult to imagine that. On top of that, when the
size of the mud is small, I can just stand as I used to, even when the mud
hits my face. I may step back a little bit though, depending on how hard
the mud hits my face. Question: what happens if someone hits my face
throwing a solid block? Am I going to be able to stand still as I could
when I was hit by mud? Understanding the two cases is what we need to
know in order to understand different types of mechanical collisions.

There are two different types of collisions that we need to focus on,
an elastic and an inelastic collision. Bonus: Most collisions are somewhere in the middle of the two types. The cases are all we need to know
as far as our understanding of the collision types, so let us go over them
one by one.

In an elastic collision, both momentum and energy are conserved,
and such is going to be the case when no energy loss is going to
occur during the collision process. This is how an elastic collision can
be defined.

We can often observe the case on a microscopic scale. You have
tons of tiny objects such as electrons moving around a nucleus. When
they collide with each other, an electron against another electron, what
is the chance of them combining and moving as a single object within
a system? Not that much. Hint: we call them fundamental particles.

In any case, they are going to most likely "elastically collide" with each other. In other words, the total size of momentum associated with individual objects is going to change but not the size of the total energy nor the total momentum within a system. Figure 29.1 illustrates the point of an elastic collision.

Whereas, in an inelastic collision, only momentum will be conserved. Energy is not going to be conserved in the collision. Let us go back to the case of the chunk of mud hitting my face. If the mud is really "muddy," then the chunk is going to hit and stick to my face, at least momentarily. I may get displaced back a bit though—following what? Yes, it is about conservation of linear momentum. This is a perfect case of an inelastic collision where the total energy is not going to be the same before and after the collision. Let us go over another example: you and your friend throw mud at each other and the mud gets combined and moves as a single macroscopic body; in this case, the size of total mechanical energy associated with the mud is not going to be the same before and after. In other words, when the total size of mass does not change but the number of either microscopic or macroscopic entities changes in the end, that is where we observe an inelastic collision. Figure 29.1 illustrates the point. Almost all the collisions we have in the real world are somewhere between a perfectly elastic and a perfectly inelastic collision. They are neither a perfectly elastic nor a perfectly inelastic collision. Therefore, in order to analyze kinematics associated with objects in a collision, you may need to first measure how elastic or inelastic the collision or the collisions are.

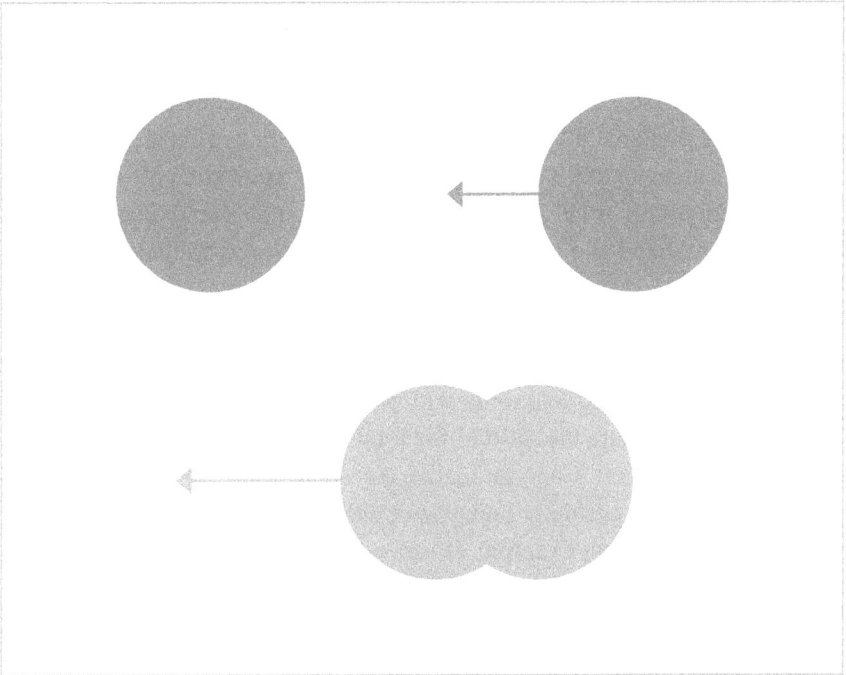

Figure 29.1: This illustrates a typical case of an inelastic collision between two objects. We have an object at rest and another one is moving toward the one that is not in motion. Once they collide, they are going to move together as a single entity. In essence, they will have the same momentum after the collision. This is a typical case of an inelastic collision, where the size of the total momentum is not going to change but the energy does. The total "mechanical" energy is going to be smaller after the collision. But we know that the size of the "total" energy within an isolated system should be the same before and after. What happened? Think about whether we can realize an inelastic collision when the two entities are test particles. In which case, can the inelastic collision be realized?

Problems:

It is hard to realize either a perfectly elastic or an inelastic collision in nature. In fact, almost all collisions on a macroscopic scale are inelastic collisions, but some collisions can be thought of as elastic collisions on a microscopic scale. Find a few examples where a perfectly elastic collision can be realized and others for a perfectly inelastic collision.

Imagine you have two of the same objects in a system. An object is in motion with a certain velocity, and the other is at rest with respect to the ground. They are going to collide with each other in a perfectly inelastic collision. Calculate the ratio of the total mechanical energy before and after the collision. Can you realize a system where no energy loss is going to occur in a perfectly inelastic or an imperfectly inelastic collision? If so, describe an example. If not, why not?

The "total" energy within a system is conserved unless an external object acts with force on one or more objects in the system. Why is the size of the total energy conserved while that of the mechanical energy is not? Describe a reason or reasons.

CHAPTER 3

Rotational motion
and other topics

We study rotational motion first. After that, we focus on understanding some similarities and differences between linear and rotational motion and review the quantities associated with rotational motion in detail. We study other useful quantities to understand the empirical aspects of the classical mechanics better too.

Day 30
Uncertainty principle

"Any physical measurement needs to be
accompanied by the uncertainty associated with it. Measurement without specifying the size of the uncertainty does not
mean much in physics."

If I am going to be interviewed by others and they ask me to choose the
single most important topic in the book, it is going to be that lesson.
This lesson is that important. Why? We need to understand the empirical aspects associated with physics in general to understand what we're
studying is all about. The bottom line is that studying physics is not just
about studying theory. We need to have some "tests" done, compare the
results to the theoretical prediction, and repeat the cycle. There are many
sources of literature where the principle is covered with more mathematical expressions. I just hope that the way the principle of uncertainty is
covered here helps you get some sense of what it is about in the scope of
undergraduate classical mechanics.

One thing that differentiates physics from pure mathematics is the
beast that we named as "uncertainty," which is associated with any type
of measurement. We "measure" quantities in physics and then compare
them to what we predicted or calculated based on theories. We then
repeat the cycle of developing or revising theories and perform the
measurements again and again, until our understanding matches with
the result that is returned from doing measurements. Physics does have
empirical aspects. It is not only about theories.

Coming back to the subject, what is the principle of uncertainty
about? Let us put it this way: measurements cannot be perfect. In other
words, we cannot measure scientific quantities with a perfect precision
anyway, either in kinematic or dynamic.

Let us go over some examples. Just as before, let us go over the case
of my moving from Charlotte to Columbia. In summary, back in 2010,
I made a trip to Charlotte, and I decided to return to my hometown by
flight, so I had to go to the airport at Charlotte. Now, let us think about

the time associated with the trip for a moment. The departure time of my returning flight to Columbia, my hometown, was at noon, but, just to be on the safe side, I left the hotel where I stayed overnight right after breakfast so that I could arrive to the airport at about 10:00 as shown in Figure 30.1. But you know what? According to my watch, I arrived at the airport at 10:00 sharp. I saw "10:00:00" on my watch. So, I claimed that I had arrived at the airport perfectly at 10 o'clock. Wait a minute. Is there something wrong with my saying that?

Let us ask a simple question: can I say that the statement above is true? Or, phrasing the question differently, can I ever say that someone or something arrives at the airport perfectly at 10 o'clock sharp? Is there such thing as arriving at a place with a perfect precision in terms of the arrival time? Answer: No, there is not. It simply cannot be with any measurement, not only in physics but also in other scientific subjects. Time cannot be measured with perfect precision; position cannot be measured with perfect precision; energy cannot be measured with perfect precision. The list goes on.

That is what uncertainty in physics is about. We simply cannot measure a quantity with 100 percent precision. Every single measurement that we are going to perform is going to have uncertainty or uncertainties associated with it.

> Neither position nor time can be mea-
> sured with 100 percent precision, and
> there is always an uncertainty or uncer-
> tainties associated with any measurement;
> thus, measurement without reporting
> the size of the uncertainty does not have
> much physical meaning or implication.

Let us think about the issue further. The bottom line is that I cannot arrive at the airport at 10 o'clock sharp. There is a "moment" at which the 10 o'clock happens in nature. However, we cannot access that moment, ever. By the same token, we cannot access a point in space. For that reason, consequently, we cannot measure any kinematical or dynamical quantities with perfect precision.

Now, let us discuss one more example that illustrates the fundamental concept of the uncertainty principle a bit more easily and clearly.

In Figure 31.1, we have two spheres: the gray one on the left shows that there is a rod that is about to go through the sphere, and the light gray one on the right is where the rod is passing through the sphere. Now, can anyone make a hole at the very "center" of the sphere perfectly using the rod? Answer: No, we cannot. We do have the center in concept, but we cannot access the center. It may sound odd, but that is how the uncertainty principle is going to come to us. Furthermore, we have multiple layers of issues that need to be addressed.

First, no one can "access" the perfect center of the sphere. The perfect center, in theory, is somewhere on the surface of the sphere for sure, and we can describe where the center is with respect to some reference position mathematically, but we cannot empirically access that point. Think about that for a while. Can you ever draw a hypothetical "point" on a surface of an object at a position with perfect precision? On top of that, can you make a rod that has the exact size that we want? Can you then identify the exact center of the rod and head that toward the center of the sphere? No one can do any of that in this world. No one.

Coming back to the main subject, in short, we do not have a way to identify and empirically measure time and position, the two important and essential quantities that describe motion in classical mechanics. Consequently, we cannot measure dynamical quantities, including energy, momentum, and force, with a perfect precision and that leads us to the uncertainty principle. Again, nothing can be measured perfectly, period.

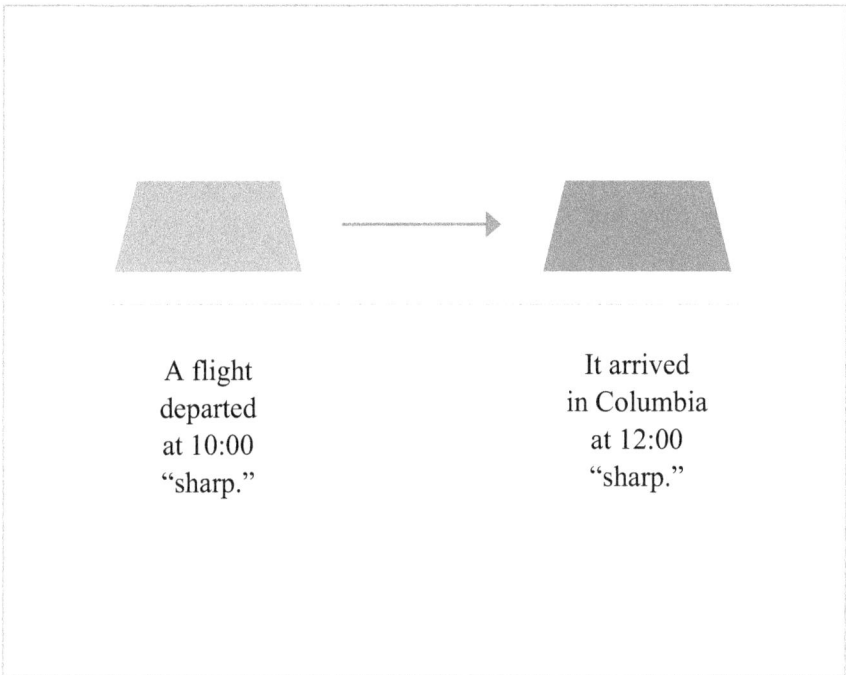

A flight
departed
at 10:00
"sharp."

It arrived
in Columbia
at 12:00
"sharp."

Figure 30.1: The flight departed from Charlotte and arrived at Columbia at 12:00 sharp in the afternoon. If the word "sharp" indicates that the flight arrived at the airport perfectly at noon, could the statement be true? If not, can we revise the statement a bit so that it does not violate the fundamentals of the uncertainty principle? Hint: nothing is perfect.

Remember: You may have heard that the uncertainty principle is something that you need to worry about only when you study quantum mechanics or advanced topics in physics. That is not entirely true; the fundamental concept that runs behind the principle applies to all areas in physics and other scientific studies, including classical mechanics, particularly when you study physics in your lab.

Problems:

Think about if you're pushing an object that was at rest and giving the object only linear motion. Write a paragraph about whether such pushing is possible or not. The object has a dimension. That means that the object that we're dealing with does has a certain size associated with it. It is not a test particle. If so, can you deliver a pure linear motion to the object or not?

Can we measure the size of either time or position with perfect precision? If not, then how are the sizes of the two correlated with each other, and what principle states the size of their correlation?

Imagine that your watch keeps to two decimals only. Calculate the size of the uncertainty that is due to the fact that your device keeps to only two decimals. How about a device that keeps to three decimals?

Day 31
Rotational motion

"Any object that occupies dimensions is
going to gain rotational motion if there is
an external force delivering an impulse to
the object. Why? We can never perfectly
target their center of mass, so our story of
rotational motion begins."

A while ago, when I was teaching undergraduate classical mechanics,
a student in the class asked me the following question: is it possible to
deliver an impulse or a force to an object so that the object attains its
motion without gaining spin? You know what? The question was a very
interesting question since a combination of a few topics in classical
mechanics needs to be addressed in order to answer the question as cor-
rectly as possible. In any case, let us not get into every single detail, but
let me give you a simple version of an answer to the question first. No,
it is not possible. The object is going to gain not only the linear motion
but also the rotational motion. Why? Let us have a look at Figure 31.1.
We have a spherically shaped object, and we want to deliver an impulse
to the object so that the sphere is going to linearly move from a place to
another place without gaining any spin. In other words, you try to hit it
with perfect precision so that it is not going to gain any spin at all but just
be in a motion linearly in physical space. However, when we try to do
so, we are going to end up with a small issue: we simply cannot access
the very center of a sphere. The center is there in theory. It does exist.
However, due to the uncertainty associated with any measurement we do,
we cannot access the point with perfect precision. We cannot get there, no
matter how many times we try.

What does that mean? What is the implication from the lesson? The
lesson is that you will always miss the center when trying to hit it with
perfect precision. No exception. You never get to the center.

You cannot deliver a force or any other dynamical quantities with perfect
precision.

What is going to be the consequence then? Anytime you deliver a force to an object, the direction of the applied force will not perfectly align with respect to that of the displacement associated with the object. In the end, the spherical object is going to gain what we called a "rotational" motion. Such is a consequence of the motion not being parallel with the direction of the applied force. In other words, the object is going to spin around the axis. For instance, you may think of a ball rolling down a hill. Figure 31.2 illustrates the rotation motion with a curved arrow. The main point is this: whenever an object interacts with another physically, two types of motion are going to occur, a linear and a rotational motion. Both are going happen, not just one.

So, coming back to you hitting a ball, the motion associated with the sphere is going to be a combination of the two: one that is parallel with respect to the direction of the applied force and one that is not. The former is what linear motion is about, and that is what we studied in the kinematics part. The latter is the "rotational" motion part. Just like the linear motion, you need to have a reference point with which the size of the motion can be quantified. For instance, the rotation is going to be described with respect to the center of the ball.

In short, following the uncertainty principle, any object is going to gain both a linear and a rotational motion when an external force is applied. In addition, a major difference between the linear and the rotational motion is that the distance to the reference position is going to stay the same in a rotational motion. Think of a ball spinning around the center. For that reason, it would be more convenient for us to introduce another coordinate system when describing rotational motion. The distance stays the same, but the displacement is changing. That is where and why we introduce physical quantities such as angle and moment of inertia. We are going to go over that later.

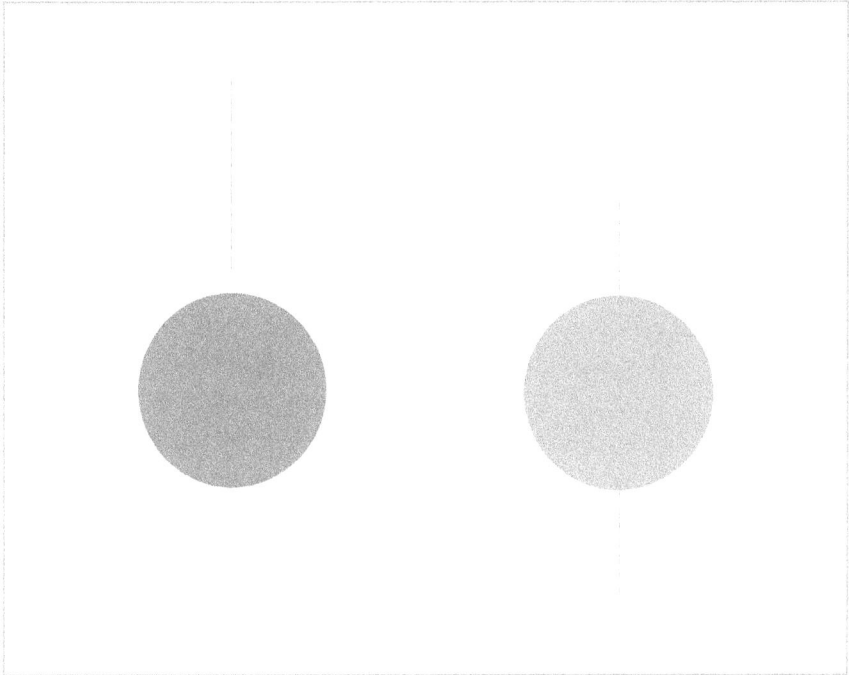

Figure 31.1: The objects in gray are a sphere and a rod that is about to go through the sphere. On the right-hand side, the rod is going through the sphere. Can anyone pierce through the sphere at the center with perfect precision if the true "center" is defined mathematically? Can anyone do that?

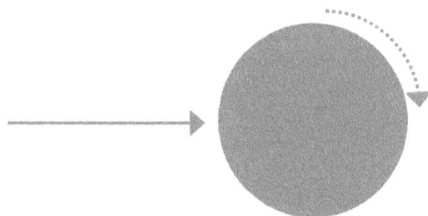

There is a point at which you can only deliver
linear motion to the object, but due to the
uncertainty principle and errors associated with any
measurement, we cannot hit the point. So, when
hitting an object with dimensions, it is going to gain
both linear and rotational motion, although the
magnitude associated with the rotational
motion could be small.

Figure 31.2: This illustrates how the uncertainty principle works. There is a hypothetical point at which force is delivered and the object with a dimension is not going to gain rotational motion. If you hit the object at that position with perfect precision, then the object is not going to gain any rotational motion, hypothetically speaking. However, that is not what is going to happen. Following the uncertainty principle, we cannot hit that hypothetical point, the center, with perfect precision. For that reason, any object with dimension in the end is going to gain some rotational motion. It could be a very small amount though.

Remember: Any mechanical motion is a combination of a linear motion and a rotational motion. We cannot realize a motion to be purely linear or purely rotational. It must be a combination of the two.

Problems:

We must introduce at least one dimension for length to introduce the rotational motion associated with an object. Think about a case where this is so and write a paragraph to describe about it.

Imagine that we have access to the very center of the sphere in Figure 31.1. Now, can we hit the very center of the sphere or not?

Day 32
Angle

> "The distance between an object and a
> reference point in rotational motion is
> going to be same, thus we need another
> quantity that lets us describe a position
> associated with an object in rotational
> motion. It is a quantity that describes a
> motion where distance is the same, but
> displacement changes as the motion goes
> on. That is where we introduce angle."

Can you define what a circle is? Mathematically, a circle can be defined as all the points whose distance from a reference is the same in two dimensional spaces. Likewise, a sphere can be defined as points whose distance from a reference is the same in three dimensional spaces. Question: how do we distinguish a point from another point that lies on a circle? Again, they all have the same distance with respect to the center, so there must be a way to differentiate one point from another. Answer: We introduce angle as another variable to describe rotational motion.

Figure 32.1 illustrates the point. We have two arrows there: the one that is drawn in the horizontal direction is a vector representing the position associated with an object in the initial state, and the one in the vertical direction is for the final state. As it moves from the initial to the final position, it has been keeping the distance with respect to the reference point the same. Question: how can we describe such a motion? How do we distinguish the initial position from the final position in the figure?

Well, we introduce angle, an important quantity that does not have a physical unit associated with it but a purely mathematical quantity being introduced to describe a rotational motion. It basically indicates how far the final position in the figure is from the initial position; thus, it is a quantity that does not have a physical unit. Here is an important point to keep in mind: the distance is going to be the same but not the displacement. In other words, the direction keeps changing as the size of the distance with respect to a reference is maintained the same while it is in motion. Question: when the distance does not change, what can we do to describe the state?

Yes, instead of specifying the position associated with an object in terms of its distance, we specify where the object is using the size of the distance vector and the angle; the size part is going to be the same, but the angle part is not while in motion. The angle changes in a rotational motion just like the way we understand displacement, velocity, and acceleration in linear motion.

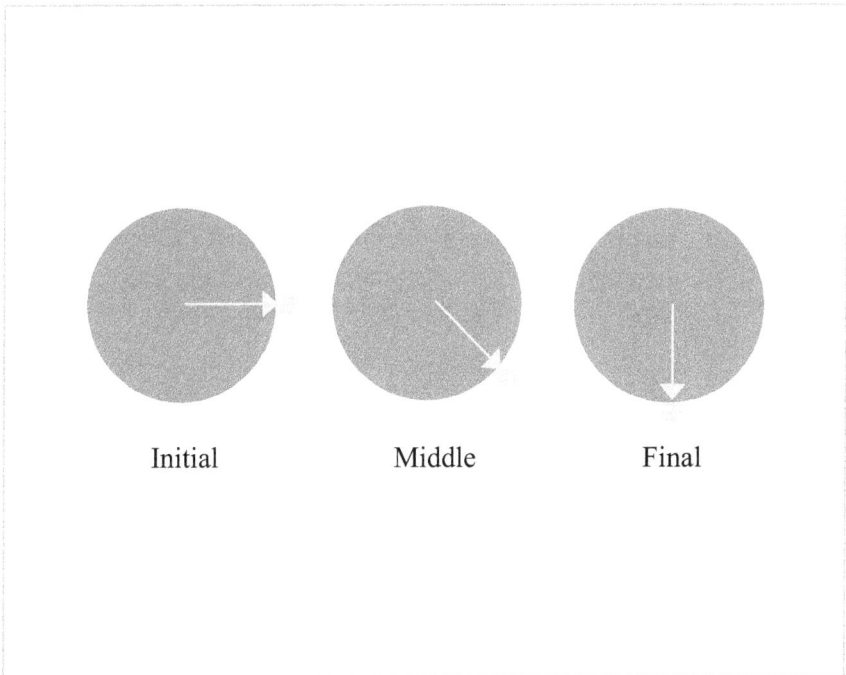

Figure 32.1: Imagine that the circle in light gray at the end of the arrow on the right- and left-hand side in the figure moved from the initial state to the middle and to the final state. You can think of this as a clock. Assume that the distance from the center of a circle in gray to the small object is going to be maintained as a constant. So, the distance does not change, but "something" changes. Here comes an interesting question: in which coordinate system is it convenient for us to analyze the motion? How about the Cartesian coordinate system in which we study vectors? But then, does the size of the arrow change as a function of time? If not, what can we do with the size, and what does that have to do with our utilizing a rotational coordinate system when analyzing the motion?

Remember: Just as in other lessons in this book, understanding what angle is in physics is also very important in order to understand the basics of what rotational motion is about. Angle does not have any physical unit associated with it. Degree or radian, the two most popular units for angle, are just the units associated with the quantity. Again, angle is a mathematical quantity, not one that carries a physical unit associated with it. In other words, to describe a physical motion, we need not only the angle but also another quantity or quantities in rotational motion. That is the moment when inertia gets introduced.

Problems:

Find the two common coordinate systems by which quantities in linear and rotational motions in classical mechanics can be used and represented. Think about which coordinate system is going to be more useful for us when quantifying that in a rotational motion and that in a linear motion.

We have two common units when quantifying the size of an angle. Find the two common units. Think about which one is more beneficial and easier to use when describing rotational motions.

Day 33
Moment of inertia

"Angle is an important quantity that is
utilized to describe rotational motion, but
it does not have any physical unit associated with it. Dimension length needs
to be introduced somewhere else when
describing a rotational motion."

Physics is fair.

Physics is really fair. Particularly, it is fair in terms of the physical
quantities that we need to deal with when studying classical mechanics.
I can tell you that for sure. If you remember that part where we were
describing "pulling ropes with less force but with more displacement,"
you may now trust what I am going to say in this lesson.

Say, for instance, you have a balloon, and you squeezed a part of
it hard with all your mighty and gusty force. Guess what is going to
happen? Yes, the other part that you could not fully squeeze is going to
get larger. In the end, the total volume of the balloon is going to stay
the same as before, unless it blows up. You might have heard that story
before, and that is what the balloon effect is about in general; if one
thing changes, something else needs to change in order to keep the total
the same as before. It may sound like what we have studied as the conservation of energy; the total energy before and after the interaction is
going to be the same before and after. In this lesson, we are not talking
about the conservation of the size of the physical quantities, but about
the "dimensions" associated with the quantities when we study physics,
especially as it describes a rotational motion. In short, the dimension gets
conserved also.

We learned that angle is an important quantity when studying rotational motions in the previous lesson. I do not question that the angle
is a useful quantity when describing a rotational motion. But you know
what? We have an issue when describing the motion with the angle and
the displacement only. What is that about? Yes, the angle does not have
any physical units associated with it. It is not a physical quantity, if you
think of it.

Angle does not have any physical units associated with it.

You may ask why we describe a motion using the angle then. We do so because the distance with respect to a reference does not change as a function of time; the distance stays the same, so that is where we are allowed not to worry about the distance part when describing the rotational motion. Do we have any problem with the distance staying as a constant?

Yes, we do. The dimension length is not a part of angle. There is no physical unit associated with angle. It is a mathematical quantity that does not carry any information regarding something associated with a physical unit. In other words, if we describe a motion using only angle, we are describing a motion without introducing the dimension length; we cannot fully describe a mechanical motion. Point: we need something else to be introduced.

Let us go back to the balloon effect. The "total" needs to stay the same before and after. If the dimension length is something we need to describe a mechanical motion, and that was introduced in linear motion but not in rotational motion, that means we need something else in a rotational motion. What is the "something else"? Where do we introduce the "something else"?

We know that mass does not change in classical mechanics. It may not be covered as a part of your lecture, but one of the critical assumptions that we need to keep in mind when studying classical mechanics is the conservation of mass. The total size of mass does not change as a function of time.

The total mass is conserved. The mass does not go anywhere.

Mass is a quantity that stays the same when in motion in general. Here comes an important point: I just mentioned that the length in rotational motion does not change as a function of time. So, why not "combine" the mass information to the length information and treat the combined quantity as a single quantity when describing a rotational motion? Now, you may know where I am heading with this. Yes, this is where we introduce the moment of inertia. For a test particle, if an object is a certain distance away from a reference and has rotational motion, the momentum of inertia is defined as mass multiplied by distance squared.

> For a test particle, if an object is a certain
> distance away from a reference point and
> has rotational motion, the momentum of
> inertia is defined as mass multiplied by
> distance squared.

You may find more mathematical details associated with the moment of inertia elsewhere, so note that the size of the moment of inertia stays the same, just like mass in a rotational motion. Again, there are quantities that do not change as a function of time, so we just want to combine them as a single quantity. That is all.

We can approach the issue associated with the moment of inertia using figures. Let us have a look at Figure 33.1. It might take a while to understand all the objects being drawn in the figure, but I think it is worth going over them one by one in order to understand the motion better. So, we have different shapes of objects present with different units. When describing linear motion in a two-dimensional space, what do we need? Yes, we need two length quantities and a mass. We need two length units since a position in two-dimensional space can be described by a combination of distances in the horizontal and vertical directions. It does not need to be orthogonal, but we need at least two of them in any case. In addition, we need a unit for mass if it is to be compared with other objects having a different mass.

Now, how about a rotational motion? Just like a linear motion, we need to introduce three quantities to fully describe the motion in space but a bit differently: one angle, one length to indicate the magnitude, and a mass. The length is needed, although it will stay the same, but it needs to be there. Can you see any other interesting points from the figure, other than the type of variables?

Let us look at the colors. You might already have noticed that the items in light gray are ones that change as a function of time and the ones in gray are not. For instance, in a linear motion, the positions in the horizontal and vertical directions change, so they are colored gray. In a rotational motion, the size does not change, so it is colored gray. Given all that, here comes an important question. If angle is something that is going to change, but the distance and the mass are not, why not combine the ones that do not change together? Answer: That is exactly where we introduce the moment of inertia. In a rotational motion, the moment of inertia is not going to change as a function of time. In short, we can treat the moment of inertia just like we do mass in a linear motion.

Remember that we do not always need to introduce the moment of inertia to describe rotational motion. It is just that introducing the quantity makes studying and understanding the rotational motion easier, so do so. We take the quantities that change as a function of time from the ones that do not change. When combining all the quantities that do not change, as colored in gray in Figure 33.1, we end up with the moment of inertia in a rotational motion.

How do we define the moment of inertia mathematically? It is mass multiplied by distance squared. Is that the end of the story? Probably not. For a test particle, it is as simple as that. If there is a point and if we are to describe its moment of inertia, that is all we need to understand. However, when calculating the size of the moment of inertia, we calculate the momentum of inertia for an individual piece of mass within the object and add them up together in the end, which is where we need to introduce calculus.

For instance, the moment of inertia for a ring with respect to a line passing the center is just mass multiplied by the radius of the ring. However, when you do the same but for a solid sphere, an individual chunk of mass within the sphere has a different distance with respect to the center of the sphere, so we need to introduce calculus to calculate the total moment of inertia associated with the shape of the object; you do the same for any other object with a different shape.

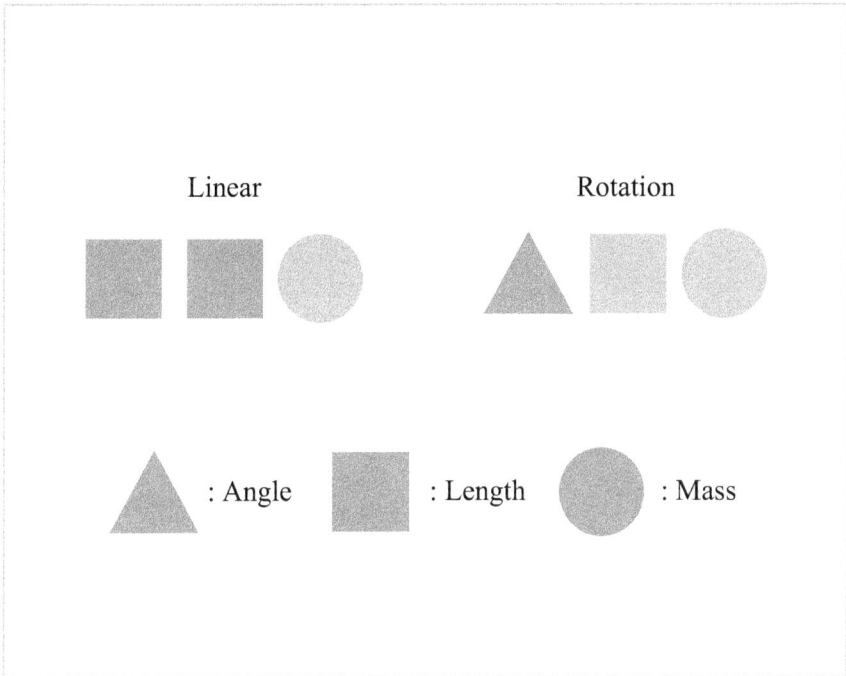

Figure 33.1: In a two-dimensional linear motion, we need two quantities in length and one in mass to describe its motion. In a rotational motion, what do we need instead?

A solid sphere is rotating with respect to an axis. The two small circles in light gray represent infinitesimally small pieces within the spere. The one farther away from the axis is rotating more in terms of its displacement per rotation, and the one closer less. How do you accommodate all that in the moment of inertia?

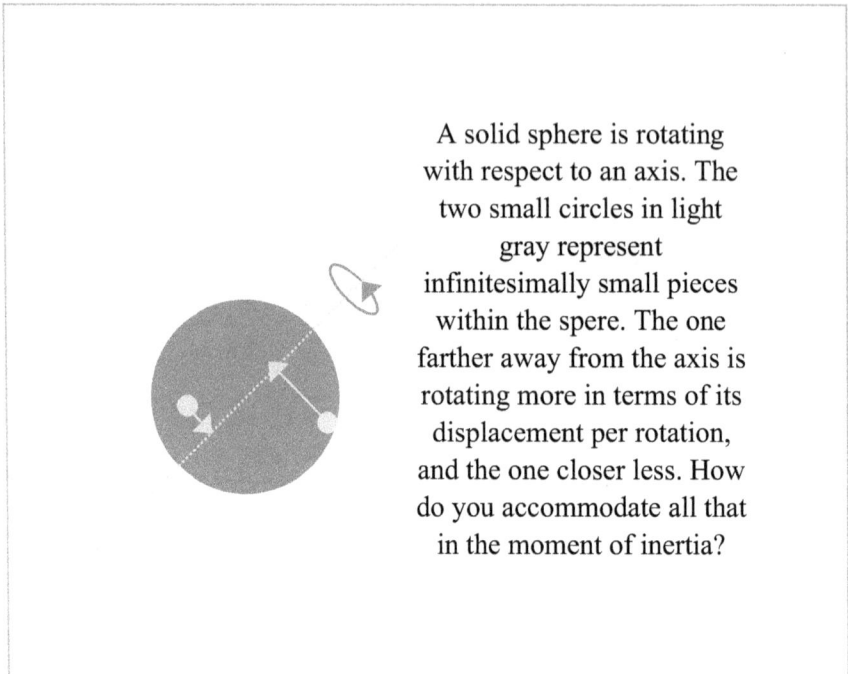

Figure 33.2: Imagine that we have a solid sphere that rotates with respect to an axis that passes through the center. The two small circles in the figure represent the two infinitesimally small pieces within the sphere. In terms of the displacement associated with the two small pieces per rotation of the sphere, the size of the displacement is going to be larger for the one that is farther from the axis, which is the one with the larger size arrow in light gray. Question: how do you accommodate for the difference in the distance to the rotational axis when calculating the size of the moment of inertia? We cannot simply assume that the moment of inertia is going to be simply the total mass times the radius squared. Why? All the small pieces shown in the figure have different sizes for distance.

Remember: physics is always fair, and it is very clearly realized when we study what the motion of inertia is about in a rotational motion. We introduced angle as a purely mathematical quantity when describing rotational motions. However, since the angle does not have any physical unit associated with it, the dimension length needs to be introduced somewhere else when describing a rotational motion. That "somewhere" is what the moment of inertia is. As a result, rotational kinetic energy and angle momentum can be described as a function of the moment of inertia.

Problems:

The moment of inertia has a dimension of mass times distance squared. It is as simple as that for a test particle. Question: can you think of a reason for our squaring the length? Why is it not just the length (instead of the length squared)?

Calculate the moment of inertia for a solid sphere with respect to a line passing through the center of the sphere. Do the same calculation for a solid disk. You may not need to do the full calculation but understand how the calculation is done utilizing calculus. Figure 33.2 is going to help you when thinking about why the size of the moment of inertia for an object with a dimension is not simply going to be the total size of mass times the radius of the sphere squared, or the radius of the circle squared. We need to do something more than that.

Day 34
Circular motion

"We can realize a uniform circular
motion by keeping the distance the same
in a rotational motion."

Question: do you know the size of the needle in your clock? If not,
why not then?

Have you ever wondered why satellites can move around Earth?
I have wondered about it from time to time—at least back in my high
school days. It may be difficult to easily show how satellites stay in space
moving around Earth with all the details. However, for the purpose of
describing their motion in terms of the mechanical energy associated with
the satellite, we can simplify the case, and that is what we are going to
have a small discussion about in this lesson.

A uniform circular motion is a special type among all the circular
motions that we can think of. It is a rotational motion where the distance between an object and a reference point is maintained as a constant and the size of the velocity associated with the object in motion
stays the same. In other words, the velocity changes but the speed
does not. A typical example that you can think of is a satellite circulat-
ing around Earth. You can think of a similar case in practice too. For
instance, you tape a piece of string to a rod, tape a tiny piece of plastic
to the other side of the string, and start rotating the string as you're
holding the rod.

Now, let us go over the motion in more detail. Why does the tiny
piece of plastic move in a circular manner continuously? What keeps
the plastic going in a rotational motion? Answer: There is a force that
balances the force that you are acting on in the system, so all the forces
acting on the plastic are going to be canceled out. That is what "centrifugal" force is. In other words, the force that you are putting on the system
as you're holding the rod is balanced out by the force that the plastic is
trying to move away from the system so that the object is going to be in a
motion where the distance is maintained as a constant while the displacement is continuously changing.

The next item that you may wonder about is how the velocity associated with the object in a rotational motion is related to the different types of forces we've described so far. Well, that is exactly where Huygens' principle is introduced. The principle says that the size of the centrifugal acceleration is proportional to the square of the velocity and inversely proportional to the size of the displacement.

> The size of the centripetal and centrifugal force are proportional to the mass of the object and the centrifugal acceleration, which is the velocity of the object squared divided by the distance from a reference to the object. The centripetal force is the force that is directed inward, and the centrifugal is directed outward.

We can write them mathematically as the following:

$$\text{Velocity} = \text{Acceleration} \times \text{Distance} \times \text{Distance}$$

in a rotational motion.

Let us now come back to the case regarding the satellites moving around Earth. As shown in Figure 34.1, we do not have a string that holds Earth in space to realize a circular motion. Therefore, we have a slightly different situation, but we can think of an analogy to the case with the rod and a plastic. I was holding the rod, but if there is no such rod in this case, what is holding Earth? Yes, it is going to be the Sun. Just like I was holding the rod so that the rotational motion could be realized, the Sun is heavily keeping the Earth in as a part of our solar system, and that is balanced out by the force that tries to make the Earth move outward when in the circular motion. In short, remember that in the circular motion there are two different types of forces, centripetal force and centrifugal force. The former is the apparent force that is pulling the object toward the rotational axis, and the latter is the inertial force. The sizes associated with the forces can be calculated by Huygens' principle.

The Earth

Centripetal   Centrifugal

So, the mechanical motion
takes place in a direction
that is perpendicular to that
of the forces.

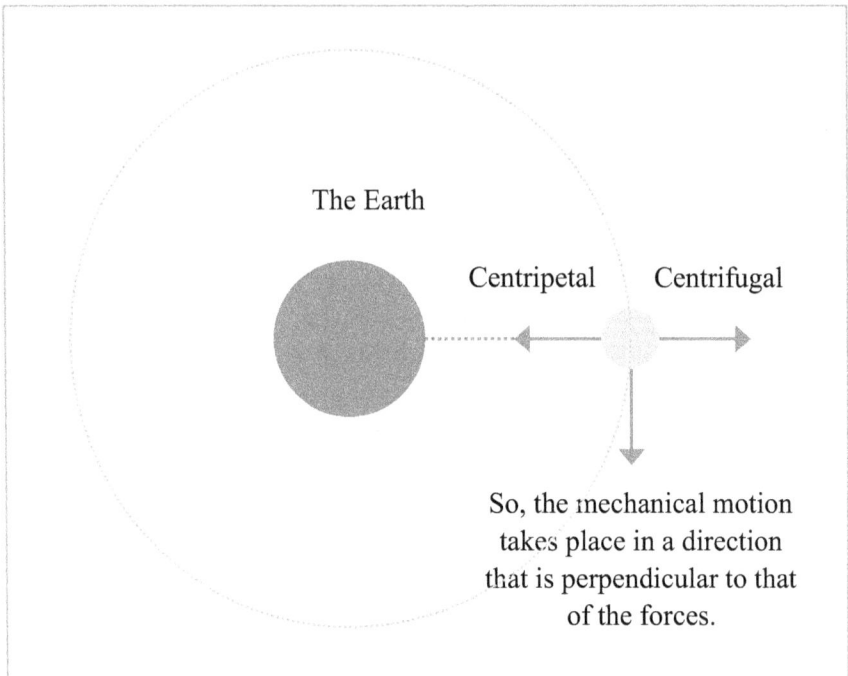

Figure 34.1: We have an object and if it happens to be in a circular
motion, we would have two forces that keep the object in motion balanced out with respect to each other. Centrifugal force is an apparent
force that pulls the object out, and that is something that we can physically observe. The other one is centripetal force, and that is what pulls
the object inward. The two forces are balanced out when an object is in
a circular motion. For that reason, the total size of the force, or the total
net force, acting on the object is going to be zero in the end. Following what? Yes, the principle of action and reaction is behind this. If the
action is the centripetal force, then the reaction is going to be the centrifugal force, and vice versa. What is the result? The object is going to
be in a motion with a constant velocity, which is exactly what a uniform
circular motion is about.

Problems:

Since the velocity associated with the object in a circular motion keeps changing, there are two different types of acceleration that we can think of: one is parallel to the direction of the motion, and another is perpendicular to the motion. Describe what these two accelerations are.

Think about what you have studied for Newton's laws and how they are related to what we have studied here as Huygens' principle. In other words, can we somehow relate Newton's laws to Huygens' principle?

Day 35
Quantities in rotational motion

"If there is a quantity in a linear motion,
there is a quantity that corresponds to the
quantity in a rotational motion."

This might be another important lesson. This lesson is not to go over a specific topic but to give you an overview describing why we study the kinematics for the cases of linear motions and why it is then followed by studying rotational motion.

In short, we have two reasons for learning the two types of motions in this order. First, it is easier to understand linear motion since it can be understood as studying motion in one-dimensional space. For instance, we have an object and the motion associated with the object is confined to one-dimensional space. On the other hand, if you think about it, we need to have a minimum of two dimensions to realize a rotational motion. Otherwise, a rotational motion cannot be realized. There is one more reason, and this might be more important than the first one and here it is: for every single physical quantity that we have covered when studying linear motion, there always is a corresponding quantity in rotational motion.

For every single physical quantity that we have covered when studying linear motion, there always is a corresponding quantity in rotational motion.

What does that mean in terms of our studying classical mechanics? Answer: We can study the rotational motion by thinking about all that we have studied in linear motion. Again, for all the quantities that we have covered in the linear motion, there are corresponding quantities in the rotational motion.

So, let us go over some important quantities. Which one can we think of easily? Yes, displacement is one of them. For displacement in linear motion, what is a corresponding quantity going to be in the rotational motion? Is it going to be the same sort of displacement in the rotational motion? No, it is not, but we have a rather different convention to describe a rotational motion. So, what can we think of? Yes, it is "angle" that is going to describe rotational motion. Angle is a corresponding quantity in rotational motion. If it is displacement in the linear motion, then it's angle in the rotational motion. The reason that the physical unit for the two motions are different is because the distance in rotational motion is maintained as a constant, so that can be pulled out as a constant. Remember: we care about what is changing as a function of time when studying classical mechanics.

We care about what is changing as a function of time.

After that, understanding velocity and acceleration follow naturally. Instead of linear velocity, we have angular velocity. Instead of linear acceleration, we have angular acceleration. You changed the word "linear" to "angular," and the list goes on for all the quantities that we have covered.

How about the dynamics part? The story is a bit different when it comes down to the work done to or by the system.

> For all the quantities we deal with in the
> rotational motion, replace the position by
> angle, velocity by angular velocity, accel-
> eration by angular acceleration, mass
> by moment of inertia, but time stays the
> same as it is.

Is that not simple? You just need to follow the rules, and you are going to get the corresponding quantities that you need when describing a rotational motion.

Remember: for all the physical quantities that we utilize to describe linear motion associated with an object, we have corresponding physical quantities in rotational motion. It is important to remember that. They just happen to look different, but they're not that different in terms of analyzing motion.

Problems:

Rotational motion could be thought of as a combination of the linear motion. That means that the kinematics equation for rotational motion can be derived from the equations for linear motions. Think about why this could be so and write a paragraph on it.

Day 36
Angular momentum and rotational energy

"It is not linear momentum, but we need angular momentum when describing a rotational motion. It is not a force but torque in a rotational motion. It is not energy associated with a linear motion, but kinetic energy associated with a rotational motion. Think about a ball rolling down a hill."

If you remember, utilizing linear momentum happened to be useful when studying linear motion; it stays the same unless an external force acts on a system. Just like that, angular momentum is very useful when describing rotational motion. The size of the total angular momentum in an isolated system stays the same unless an external entity acts on it. As described in a previous lesson, we do see similarities between quantities in linear and rotational motions.

Linear momentum is defined as the mass of an object times the velocity of the object. So, it is a vector. But we have a slightly different situation for angular momentum. Since the angle does not have any physical dimension associated with, angular momentum cannot be defined the same way as linear momentum. Question: what do we need to do? Answer: We need to bring the dimension length back into the definition. For that reason, angular momentum is defined as the mass associated with an object times the cross-product of the velocity and the position vector associated with the object. Why is this important? Just like the linear momentum in linear motion, angular momentum is conserved in an isolated system. On top of that, some intrinsic properties associated with momentum in both type of motions have similarities, and we are going to go over them later.

So, if angular momentum is a counterpart for linear momentum, then how about energy in a rotational motion? Yes, there is energy associated with rotational motion, just like energy associated with linear motion. In a linear motion, we have two different types of energy, kinetic and potential energy. The former is a quantity that has to do with an object being in motion, and the latter has to do with the distance and something to be realized later in other forms of energy. Again, the potential energy has something to do with the distance and is not associated with motion, and it is going to be the same for both the linear and the rotational motion.

But it is for kinetic energy that we need to differentiate the two different types of the motions. In terms of linear velocity, the kinetic energy happens to be proportional to velocity squared. In terms of angular velocity and other quantities in rotational motion, the same energy can be defined by what? Yes, it is going to be defined in terms of the moment of inertia. It is going to be proportional to the size of the moment of inertia and the size of the angular velocity squared. Remember: All you need to do when studying rotational motion in classical mechanics is replace the variables that you have studied in linear motion with those for rotational motions.

How about a quantity that corresponds to "work done" in a rotational motion? Can we realize the "work done" in a system when having a rotational motion? We may not, and that is because the part of displacement that is parallel to the direction of the force only matters when it comes down to the work done by or to a system. However, in a rotational motion, we do not have such a component, if you remember. The component of the displacement that is "not parallel" to the force is all we have in a rotational motion. Going back to one of the previous lessons, you are going to hold a string hard to keep a piece of plastic in a rotation motion; the direction associated with the force is toward the center of the motion, whereas the direction of the displacement is perpendicular to that of the force. Point: we cannot realize "work done" in a rotational motion. That is just part of the definition. At the same time, the part that is not realized as a part of work done needs to be addressed, and that is exactly what "torque" is in a rotational motion. You take the force and the component of the displacement not parallel to the force and take what we call "cross-product" of the two in mathematics, and that will give us the size of the torque.

The component that is parallel to the direction of the force contributes to the work done to or by the system, whereas that which is not parallel to that of the force goes down to torque in a rotational motion.

Figures 36.1 and 36.2 illustrate the points well. In any case, you may want to think about opening a door. Imagine you have a large door, and you push the door in a way that the direction of the force is parallel to the direction of the door. Can you push the door to get it open or not? If you can, I may need to have a chat with you because I tried to do so, and I could not get it open no matter how many times I tried it.

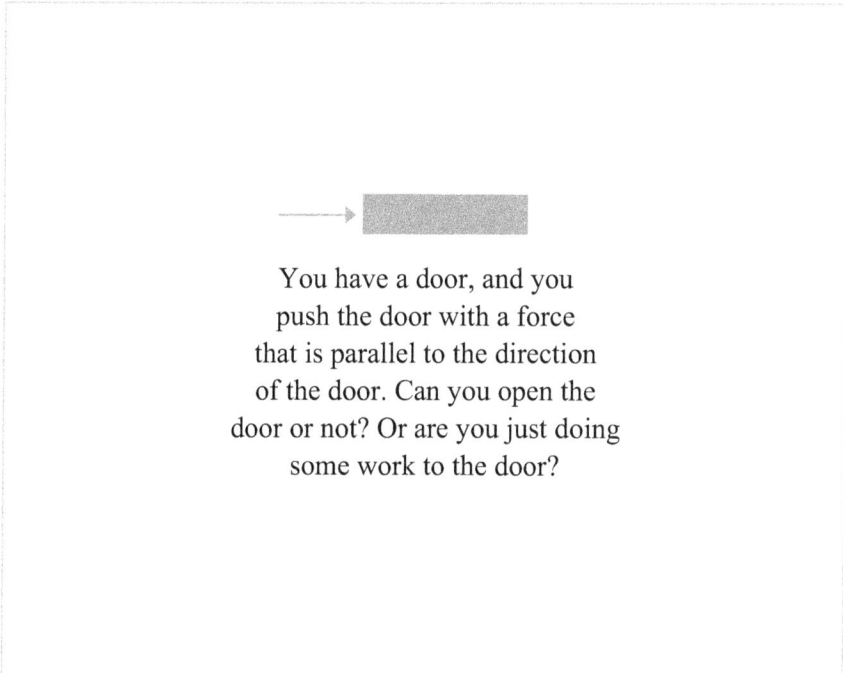

You have a door, and you
push the door with a force
that is parallel to the direction
of the door. Can you open the
door or not? Or are you just doing
some work to the door?

Figure 36.1: Imagine that you have a door, and you push it in a way that the direction of the force being applied is parallel to that of the door. Question: can you open the door or not?

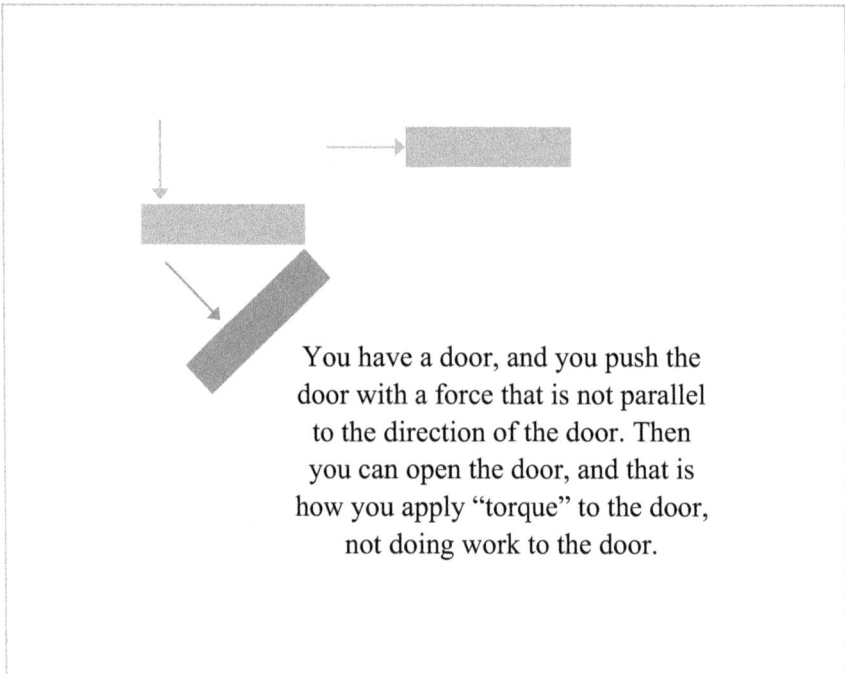

You have a door, and you push the door with a force that is not parallel to the direction of the door. Then you can open the door, and that is how you apply "torque" to the door, not doing work to the door.

Figure 36.2: If you push the door parallel to the direction of the door, you are not going to be able to open it. Instead, you need to push it from the front, as illustrated on the left side of the figure. That way the direction of the force is not parallel to the direction of the door, and the door is going to start rotating with respect to its axis and you are going to open the door. That is how you apply torque to the object, not do work to the object.

Remember: if you have studied the basics of "cross-product" in mathematics, it might be easier to understand the direction associated with the torque and the size of it. If you have not done so, it is okay. Just keep in mind that the component that is not parallel with respect to the direction associated with the object to be in motion is going to make contributions in a rotational motion.

Problems:

Linear motion is a specific type of rotational motion. Think about why this is so and derive the definition of linear momentum from the definition of angular momentum. You may think of an object rotating with respect to a reference point and the distance between the object and the point is relatively very large.

We have no such quantity as rotational potential energy, but there is rotational kinetic energy. How come this happens?

Day 37
Kinetic theory

"Imagine that we have a million objects
to deal with in a system. We need a
theory other than what we have studied
before. We cannot deal with them individually. We need to model it differently
and think of the case differently."

Let us go back to my moving from Charlotte to Columbia. In essence,
what we had studied was about analyzing someone moving from one
place to another place using kinematical quantities such as velocity and
displacement.

Now, let us make the case a little bit more interesting than a single
person moving from Charlotte to Columbia. Imagine there happens to
be bad weather forecasted in Charlotte, and more than a million people began to move down to Columbia from Charlotte temporarily, by
driving their cars or by taking a plane or by any other means. Question:
can we analyze the motions associated with the million people individually? Can we precisely analyze their motions if they happen to move all
together like that? In other words, can we analyze the motion of every
single object in a system when there are millions of such objects present?

Answer: Yes, we can, but hypothetically speaking. We just need
to apply all the physics laws that we have studied so far. However, is
it worth doing so? Most of all, is it really going to be accurate enough
based on our calculations to be applied in a practical case like that? Well,
it may not be so. Even analyzing motions associated with three objects in
a system, just adding one more in a system where we have two objects,
could be a very demanding task. You may need a very powerful comput-
ing resource to be precise when analyzing their motions, and you may
still end up not being able to quantitatively analyze the motions with a
reasonable precision. That is how demanding it is—and imagine you
have a million.

So, here is what we are going to do: we model the motion associated
with the million people "as a whole." In other words, we do not analyze
the motion associated with the individual entities but try to analyze what

the outcome is going to be when we treat all the entities as a single entity. Imagine that you saw a picture of a galaxy published in an astronomy journal. How many of you are going to be able to count all the planets within the galaxy? Can you even count them?

No, you may not, and that is one of the reasons for introducing kinetic theory. In kinetic theory, we analyze the motion of the entities within the system as a whole. We do not mind what is happening to the individual entities too much. We analyze the motion associated with the multiple objects in a system instead. Then we tried to extract some useful dynamical quantities from analyzing their motion.

> In classical mechanics and thermodynam-
> ics, kinetic theory is about studying how
> a system behaves as a whole in terms of
> their kinematical and dynamical quan-
> tities associated with all the entities in
> the system and studying their outcome.
> We do not study that on the individual
> entity level.

Let's think about another case. Imagine there is a plastic bag, and there are tons of atoms within the bag, as illustrated in Figure 37.1. The left side of the figure is atoms within the bag that are in motion. Each individual atom has its own velocity and acceleration following what we learned in kinematics and dynamics, but there are tons of atoms like that within the bag. That means that their kinematical quantities change as they pull each other in gravitational interaction. There are many different combinations that need to be considered, and it is almost not worth doing so. However, what is worth doing is illustrated as gray in Figure 37.1. There, we study the kinematics of all the entities in the bag and approximately calculate the size of the kinetic energy associated with all the atoms within the bag. Again, when we have more than two objects in a system, analyzing the motion associated with individual objects is not that realistic. We need something new. By the way, this is where we introduce the concept of the center of mass, and we are going to go over it later.

In short, if there are too many objects within a system, we need to simplify it, and that is where the kinetic theory comes in and that is a basis for thermodynamics in physics. It would be great if you can model them individually, but if not, that is okay. We will treat them as a group.

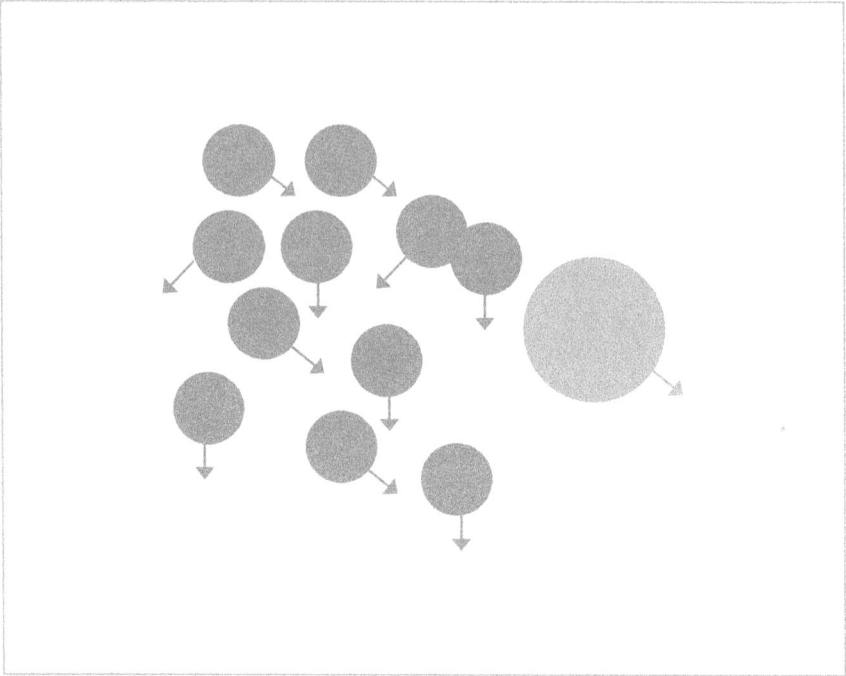

Figure 37.1: We have many objects with different masses, velocities, and accelerations. If it is in the order of 100, we can measure them and calculate the total. But if we have objects in the order of a million, it is going to be quite a different task to measure the kinematical quantity associated with individual objects and add them together. That is where we have kinetic theory; kinetic theory is about estimating how multiple objects' kinematics and dynamics are going to be translated on macroscopic scale. Temperature is a good example; we cannot measure the motion of every single molecule within a space, but by measuring the change in temperature, we know how the individual molecule's behavior is manifested in the macroscopic world.

Problems:

Thermodynamics is a branch of physics where the kinetic theory is taken as a basis. Find three laws in the introduction to thermodynamics and write a paragraph about how the laws are relevant to the fundamentals of kinetic theories.

Imagine that the circles in dark gray on the left-hand side of Figure 37.1 are placed in a box and the box as a whole happens to be at rest on the ground. Calculate the total size of the kinetic energy associated with the box and that of the potential energy with respect to the ground level. Is the total size of the kinetic energy zero or not? If so, why so? If not, why not? If not, then how is the size of the kinetic energy going to be represented? Hint: search for a first law in thermodynamics.

Day 38
Test object and center of mass

"We need to introduce the center of mass
when we deal with objects having structures in space. If no dimension is introduced, then we deal with test particles or
test objects."

Most of us live on the surface of Earth. Some of us could be in the air for
many hours, either flying or parachuting, but most of us spend most of
our time on the surface. In other words, we are in contact with the surface
of Earth. Figure 38.1 illustrates the point. The gray circle is the Earth, and
the light gray one is the people residing on it in the figure. Here comes an
interesting question: is there potential energy between the people residing
on the surface and the Earth? Do we have potential energy with respect to
the Earth even when we are in contact with the Earth? Why? Should there
be no potential energy at all in the system?

You may think that there is no such thing as potential energy between
people and the Earth since we are not at a distance with respect to the surface. You may think so since we are "in contact" with the ground surface.

But you know what? We are. We, as entities in physics, are in fact
"at a distance" with respect to Earth, at least when it comes down to analyzing the case in physics, even though the two appear to be in contact
with Earth. Why? Because we need to understand the state associated
with objects in classical mechanics in terms of what we call the center
of mass. When we calculate some physical quantities, we calculate them
in terms of the center, not in terms of some random position within the
object. In other words, you have an object in space, and you want a point
in space to represent the object. How do we do that? Yes, there we introduce the notation of the center of mass.

The center of mass is a vector representing a position at which we
can think of all the masses of an object or objects are being concentrated
into. In other words, there is a hypothetical point in space, and the point
is where all the mass associated with the object is thrown into.

It might be easier to understand the mysterious "center of mass"
by going over some figures. Let us start with Figure 38.2. The apparent

distance between the object in gray and that in light gray is shown as an arrow. Again, it is from the surface of an object to that of another object. Therefore, when they are in contact with respect to each other, then there is no apparent distance between the two. Question: is that the distance that we need to use when analyzing their state in physics? No, it is not. We need a distance that goes from the center of the mass of an object to that of the other one. Why? Because we need to measure the distance with respect to the center of mass for each object. We need a point representing the object with structures in space. Again, it is not the distance from a surface of an object to that of the other but the distance with respect to their centers of mass when we analyze the system.

Coming back to the definition, Figure 38.3 illustrates the point further. The arrow runs from the center of the circle on the left to that on the right. Question: how should we calculate where the center of mass is? How should we locate it? Answer: You find a "weighted" position. In other words, the center of mass weighted by the whole mass associated with the object is the same as that weighted by infinitesimal pieces of mass within the object. For instance, the center of mass associated with a sphere is the center of the sphere. It may not be that difficult to imagine since it is symmetrically shaped. For a rod with a uniform density, it is going to be the middle of the rod where the center of mass is located. You may find the corresponding mathematical expressions in other literature.

Let us go over a dynamical case. We have an object in gray and another in light gray in Figure 38.3, and they both are moving down straight with the same velocity. Now, here is an interesting question: can we somehow "combine" the two objects as a single object and treat the single object the same as the system of the two objects? In other words, can we treat the two objects as a single object as represented by their center of mass? Answer: Yes, we can, and that is the merit of introducing the concept. As shown in Figure 38.4, the object in light gray is the center of mass representing the mass of the two objects being all concentrated and that red "point," which is referred as a "test particle," a hypothetical point, can be thought of as the same as the combination of the two objects; we can analyze the system using the center of mass into where the sum of the two masses is being concentrated. For a single object, as illustrated in Figure 38.5, the whole system of objects can be treated as a hypothetical point in space, a point colored as gray on the right-hand side. Again, the center of mass is just a point weighted by the entire mass of an object or objects within a system, and it is a position vector to which we can measure kinematical quantities with respect to other objects. You deal with the point instead of the object with structures.

Now, if you read through this lesson carefully, you may have the following question: how do we represent a position within the object, a position that is not the center? For instance, if you have a sphere and you want to mathematically represent a point in the object but not the center, how do we do it? How do we represent a single point that is about half of the radius away from the center? Answer: We use another vector that is representing the position with respect to the center of mass. In other words, you have two vectors to represent the position, one that goes from the reference to the center and another from the center to the position that you are interested in. What is interesting, though, is that you may not need the two vectors, but just one instead, in cases where the whole body has the same kinematical quantity. Think about that.

In short, the center of mass is introduced to represent the whole body in space as a point, a point into which the mass of the whole body is thrown, and we analyze the system using the position and other quantities associated with the point. The hypothetical object represented as a point in space is what we call a test particle.

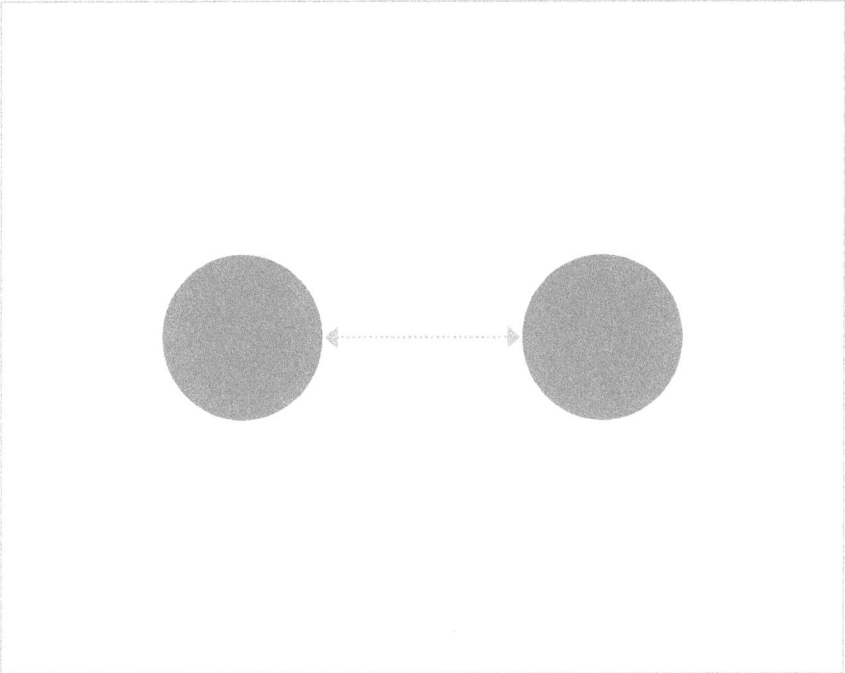

Figure 38.1: The apparent distance between the two objects is represented by an arrow, the arrow that is drawn from the surface of one
object to that of the other. However, all objects have a certain dimension
in space; they occupy space, practically speaking. Therefore, the distance that we need to think about when trying to understand and analyze
the physics running behind is to use the distance that goes to the center
of mass, not to the surface. Is it going to be the apparent distance, the
distance with respect to the surface? Yes, it needs to be the latter.

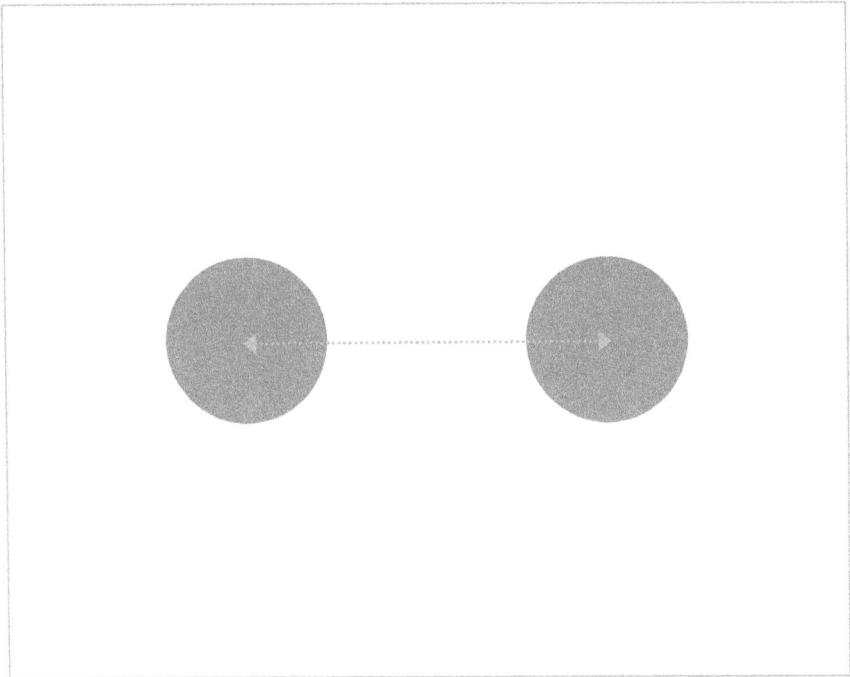

Figure 38.2: We need to introduce the concept of the center of mass when studying classical mechanics. The distance between the objects that have certain dimensions associated with them needs to be measured with respect to the center of mass for each object. The distance between the two circles in the figure is represented by the arrow since the center of mass associated with a circle is the center of the circle, not the edge that is nearest to the other object. For instance, the distance between you and Earth is going to be about the radius of Earth. Why is this so? Why is it going to be important when it comes down to measuring the size of the acceleration of you by the Earth?

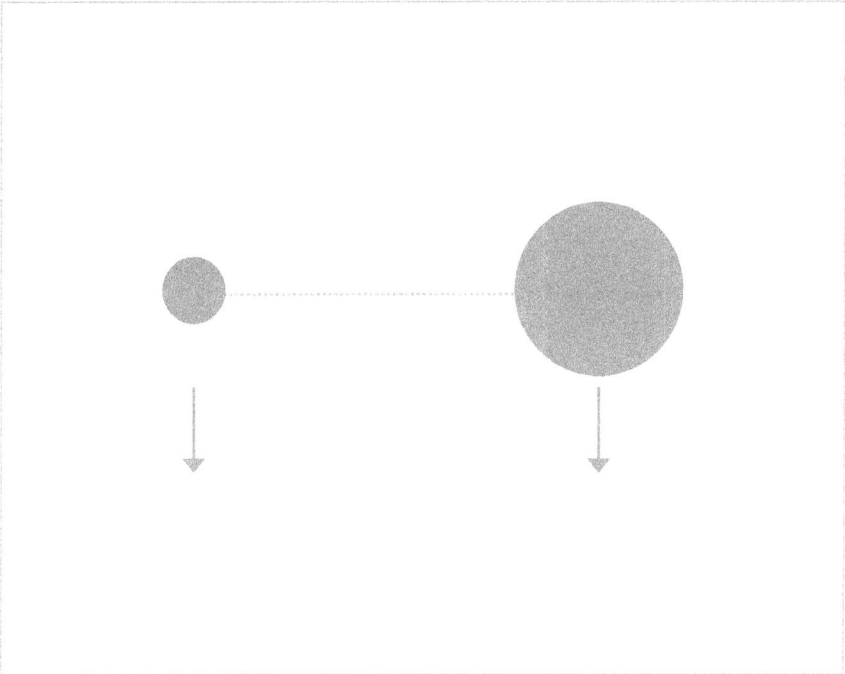

Figure 38.3: We learned that the center of mass for a single circle is the center of the circle. Now, what if we have more than one object in our system and we want to calculate the center of mass for the entire system? We have multiple objects in a system. For instance, when we have two objects as shown above, how do we calculate the center of mass representing the system of the two bodies?

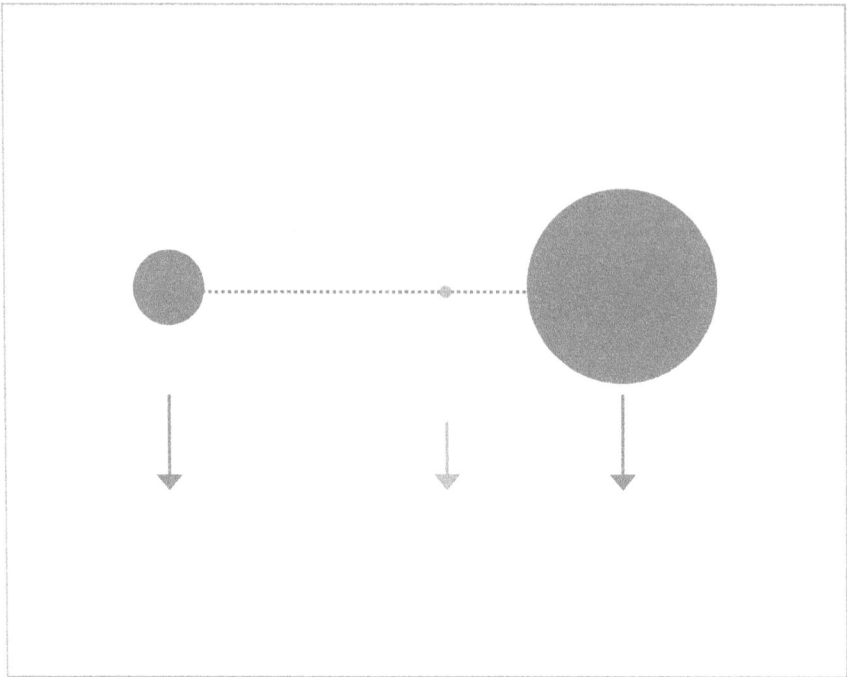

Figure 38.4: The center of mass is a position vector that is weighted by the entire mass of the system, and it is the same as the sum of that of the infinitesimal piece within the system. Therefore, when we have two objects, we simply weight the mass associated with them by their distance vector with respect to some reference. If you do so, then the light gray point in the figure is what we end up with, and that is where the center of mass is. We can think of the two objects moving down straight as the same as the light gray point moving down with some velocity. The motion associated with the two objects can be represented by that of the test particle colored in light gray.

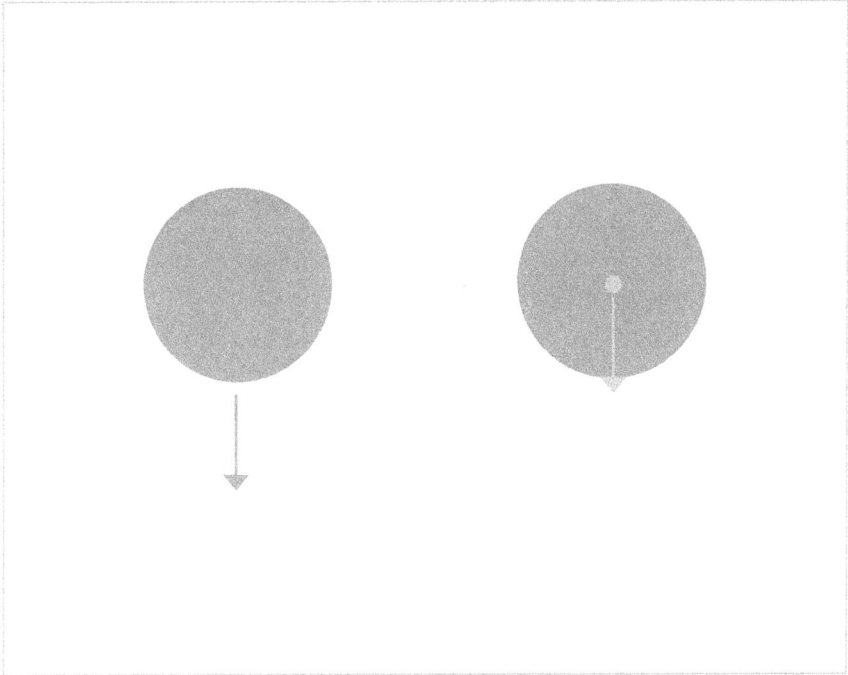

Figure 38.5: This illustrates an object in motion and where its center of mass is going to be. The center of mass happens to be the center of the circle, and we can think of the center of mass as a position into where all the mass of the object is going to be concentrated. In other words, the gray circle on the left side moving with a certain velocity can be represented by the test particle colored light gray on the right side in motion.

Problems:

The center of mass has an implication when analyzing and studying the rotational motion associated with an object. Think about why that is and describe why it is so.

When we calculate the acceleration caused by Earth on you, we use the center of mass associated with Earth. You may do the same but for that caused by you on Earth. When you do so, and if you add them up, the sum is going to be almost the same as that of the Earth on you. Think about why this is so.

Imagine that you want to analyze the motion associated with an infinitesimal piece of mass within a circle on the right-hand side in Figure 38.5. Do you think that analyzing the motion of the center of the mass is going to be sufficient to understand the motion associated with the small pieces? If not, what do we need to do to analyze the motion in terms of the motion of the center of mass? What "else" do you think we need to introduce to fully describe the motion associated with the object?

Day 39
Universal law of gravitation

"The universal law of gravitation is one
from which we can calculate the size
of gravitational interaction. It is neither
Newton's laws nor Huygens' principle."

Newton's laws are about motion. They are not about gravitational inter-action in general.

Let us go over a story that is somewhat different from the ones introduced in previous lessons. Imagine that you have two of the same magnets. Each magnet is about 0.1 kg, and you are going to momentarily place them on a table, keeping them about 1 cm apart from each other. Question: what is going to happen? Answer: Yes, they are going to pull each other. They are going to be in motion toward each other because of their having an electromagnetic interaction. Object 1 is going to move toward where object 2 is located and vice versa. The force created by the magnet is an attractive force.

Now, let us think about a similar situation but a slightly different one. Imagine that you have two pieces of plastic; they are the same shape as the magnets and have the same mass. You place them on a table just as you did with the magnets. Question: can you guess what is going to happen to the two pieces of plastic? Answer: They are not going to move; you are most likely not going to observe displacement on a macroscopic scale. Why is that? Because the gravitational pull between the two pieces of plastic is not going to be strong enough to overcome the friction between the plastic and the surface of the table, whereas the electromagnetic force between the two magnets most likely does. What is my point? The point is this: we can measure how strong or weak the gravitational interaction between two pieces of plastic is, following the universal gravitational law. It is by using the gravitational law that we can calculate the strength of the interaction. Remember, it is the universal gravitational law, not Newton's laws.

Why not Newton's laws? The laws tell us the size of the force with respect to mass and acceleration, so you may think the laws are suffi-cient for us to calculate the strength, but in fact it is not. Newton's laws describe mechanical motions in general. In other words, if the position associated with an object changes over time, then we use the law to analyze the case. The law holds no matter by which external force a motion will occur. There, following Newton's first law, we introduce the "inertial" mass associated with an object. The mass we deal with in Newtonian dynamics is that mass, not the mass that can be calculated from their having a gravitational interaction.

That brings us to an even more subtle and important question: how do we know that the acceleration of the pieces of plastic is due to the gravitational interaction? We do not. In other words, the mass that can be calculated from Newton's laws is not necessarily the same as that from the universal law of gravitation. The one that is calculated from the former is what we call "inertial mass," and the one from the latter is "gravitational mass." In short, it is universal gravitational law that tells us the strength of the interaction, not Newton's laws.

Remember: understanding why we need the universal gravitational law instead of Newton's laws when describing gravitational interaction is a bit of a subtle area, so you may need to read this lesson a few times to understand and grasp the core concept. This lesson is something that I am encouraging students taking undergraduate classical mechanics to think about.

Problems:

Find and read literature that describes the difference between inertial and gravitational mass and describe the difference between them. You may need to read Newton's laws and the universal law of gravitation first.

Derive the universal gravitational law from Newton's laws. You may need to refer to some literature published in journals. Feel free to assume and start with a predefined functional form.

Day 40
Introduction to classical mechanics in one hour

"If objects have mass, then they interact
gravitationally. We analyze the cause and
the effect, and that is what we study here
in classical mechanics."

Yes, we all do; we all want to understand the entire scope of classical
mechanics in a day or so. If such could be realized, that would be so nice,
wouldn't it? In fact, we may not need to spend an entire day to understand the entire scope of classical mechanics—but just about an hour.
How nice would that be?

Well, sadly, the truth is that you may need to spend a reasonable
number of hours to understand the scope. But how about we just go over
the basics, so we will have a less hard time studying for it later, particularly when you are preparing for some sort of test in the future.

If you happened to be one of those raising such questions or similar,
then this lesson might be a good fit for you. It could be true that some of
you do not have time to read a whole book these days, particularly if you
are one of those students preparing for a medical school admission test
or something similar. Then this lesson is something that you really want
to breezily read to grasp what classical mechanics is about in an hour.
If you are a high school student trying to get some sense of what classical
mechanics is about in a day or so, this lesson is going to be a good fit too.

Nonetheless, it would be nice if you could go over everything covered in this book, at least once, before taking an introduction to classical
mechanics course or when you're concurrently taking the course. So,
for those of you who have more than a day to spare, I encourage you to
read this whole book, or at least a few of the most important lessons,
especially if you had a hard time understanding the mathematical portion
covered in undergraduate classical mechanics.

If not, then just go over the figures in this lesson. The figures are not
to cover the entire scope of the classical mechanics but to cover some
that lead you to understand the rest of the materials with fewer barriers.
Specifically, we are going to discuss why we need to study mechanics

and the need to bring gravitational interaction into classical mechanics. Why do we study classical mechanics, and how do we need to approach the subject? This lesson has mostly figures as opposed to written sentences. In other words, there are just a few figures that you need to go over, and read the captions for the figures. That is all that I am going to ask you to do. It is not going to take more than an hour of your time, so let us begin.

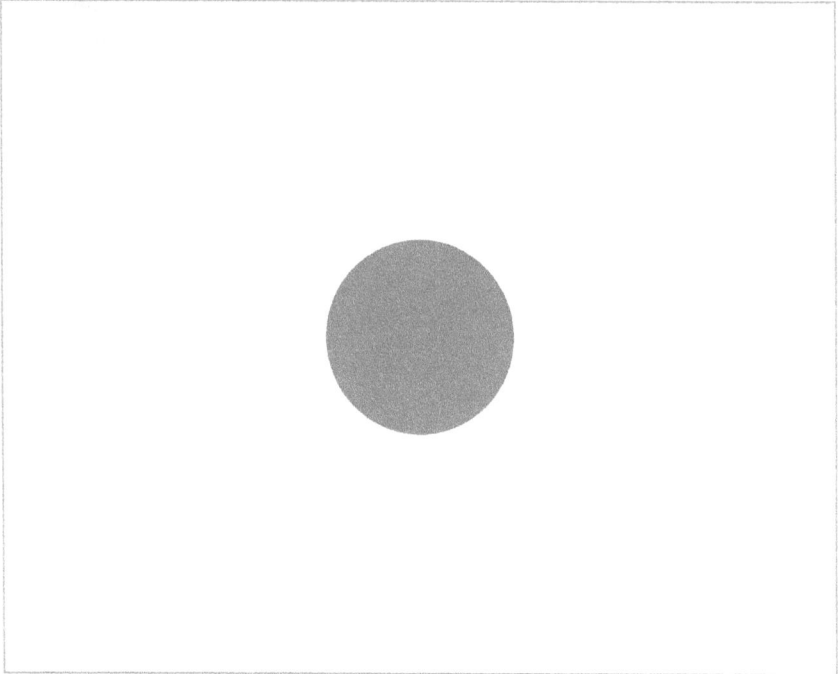

Figure 40.1: Let us look at the figure. We are going to start with a simple case, maybe the simplest case that I can think of. We have a single object in the center. It has a spherical shape. Here, what is important is that this is the only object we have in a system, and the object happens to have no mechanical motion associated with it. In other words, the object is at rest with respect to a reference, thus the position associated with the object is not going to change as a function of time. Question: is there something we can do in terms of analyzing its motion? Probably not. Why? It simply is not moving. The position associated with the object is not changing, so there is no quantitative or qualitative analysis that we can perform; there is no point in studying classical mechanics at all if this is all we need to deal with. Just remember that everything that we do and study in classical mechanics is about motion. In classical mechanics, position gets changed as a function of time, and we introduce other useful quantities such as velocity and acceleration and study the motion to do some prediction in essence.

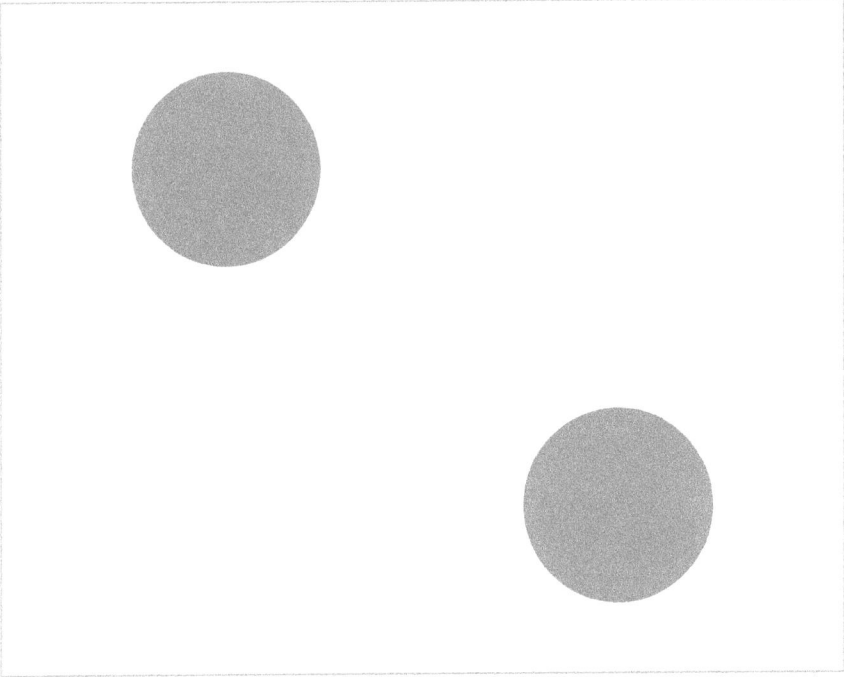

Figure 40.2: Well, the previous figure was not that fun, although there is some implication to it. If you are interested what implication I am talking about, you may refer to lessons that deal with Newton's laws. Coming back to the story, we had only one object, and it was not moving, so that was the end of it. Here, we introduce one more object, and this is where things get more interesting. We have two spheres in a system. There was one, but we introduced one more; thus, we have two objects in total. Here comes an important question: what is going to happen to the position associated with the objects with respect to time? It does not matter whether the object was in motion or not, but assuming that the two objects we have in the figure were at rest to begin with, what is going to happen right after we place them in a space like that? Answer: They are going to pull each other. Any object with mass is going to pull another object with mass in a space, and that is how they are interacting gravitationally. In other words, the object on the top left is going to start moving toward the object on the bottom right and vice versa. Remember: when we have more than one object, they are going to have some influence on the motion of one another. If you are wondering what that is about in more detail, you are ready to read the dynamics chapter in this book.

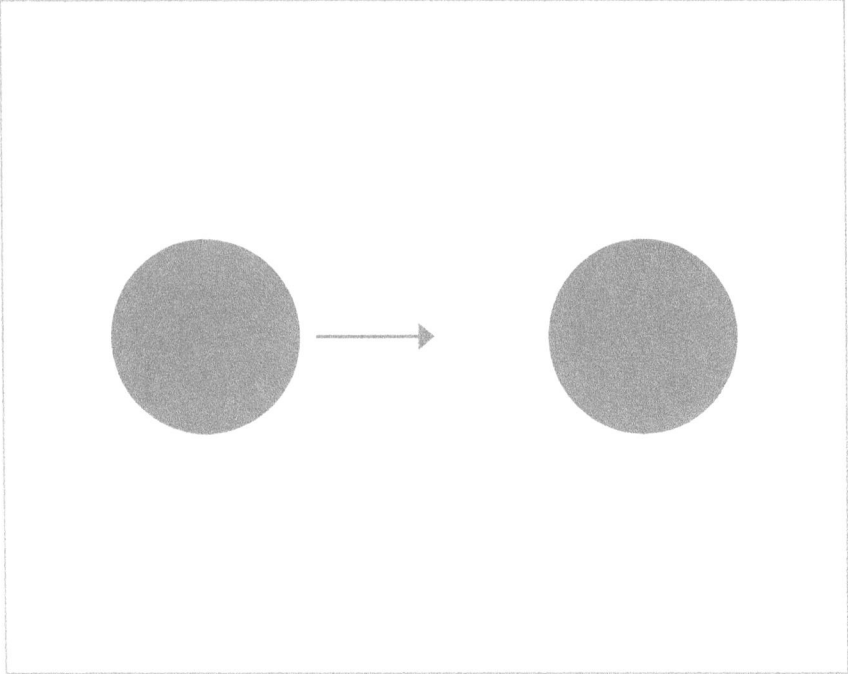

Figure 40.3: Here we are going to show the two objects in the previous figure in the horizontal direction. The object on the left is going to move toward the object on the right. The object on the right is the "cause" of the motion associated with the object on the left and vice versa. The object on the right is going to start moving toward the object on the left. Remember, they are going to pull each other because of the gravitational force between them. If there were a single object, it would not move unless it somehow gained its motion from somewhere.

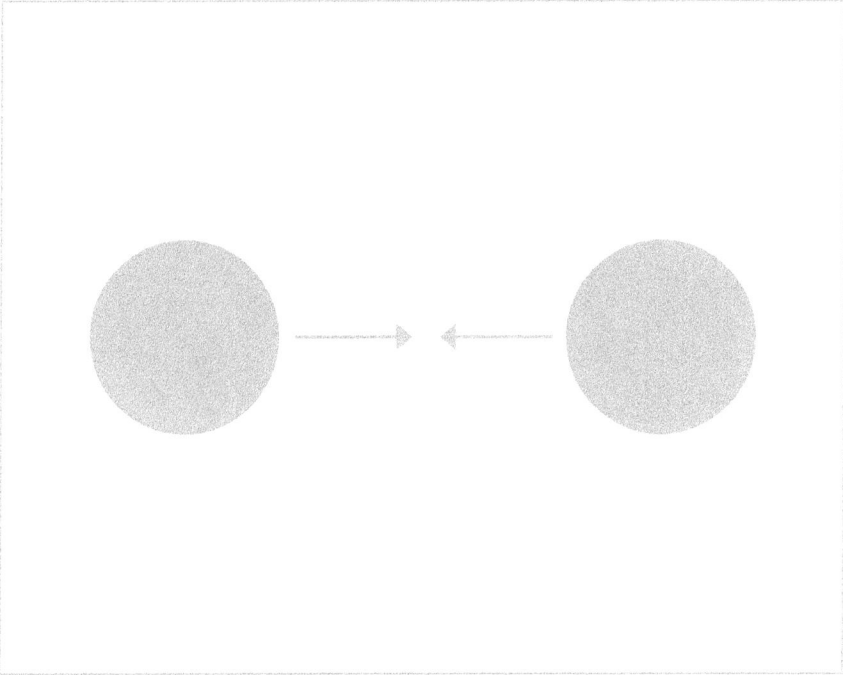

Figure 40.4: This shows that the object on the left is going to move to the right and the object on the right is going to move to the left. They pull each other. Question: how strong will their pulling be? Answer: The size of their pulling is proportional to the mass of the individual object and inversely proportional to the distance squared between the two. Did you hear about the universal law of gravitation? This is what that is. Due to their gravitational interaction with each other, things move, and that is what we study in the kinematics part of mechanics. You may wonder about the size of the arrows in the figure. They are the same length, and that is to indicate that the degree of their pulling with respect to each other is the same but in the opposite direction. It is one of Newton's laws, and if you are wondering what that is about in more detail, it is time for you to read Newton's laws in the dynamics chapter.

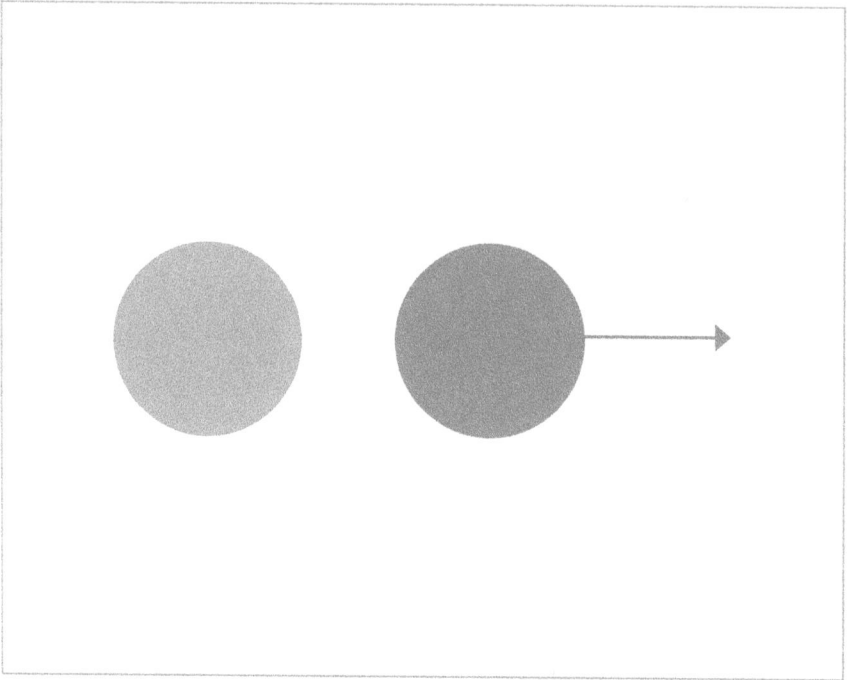

Figure 40.5: This shows the object on the left side in the previous figure. The grey circle color is the object at the beginning, and the one with the light grey color is where the object is after a certain time is elapsed. The object is going to be "in motion"; it is going to move toward where another object is in a system. The position associated with the object changes as a function of time. In other words, the object is going to be displaced in space, and that is where we introduce the notation of the distance and the displacement. In practice, the size of the displacement and time is going to be different, and we want to distinguish them. How? That is where we introduce quantities such as velocity and acceleration. If you are wondering what all of those quantities are about, it is time to read about them in the kinematics part of this book.

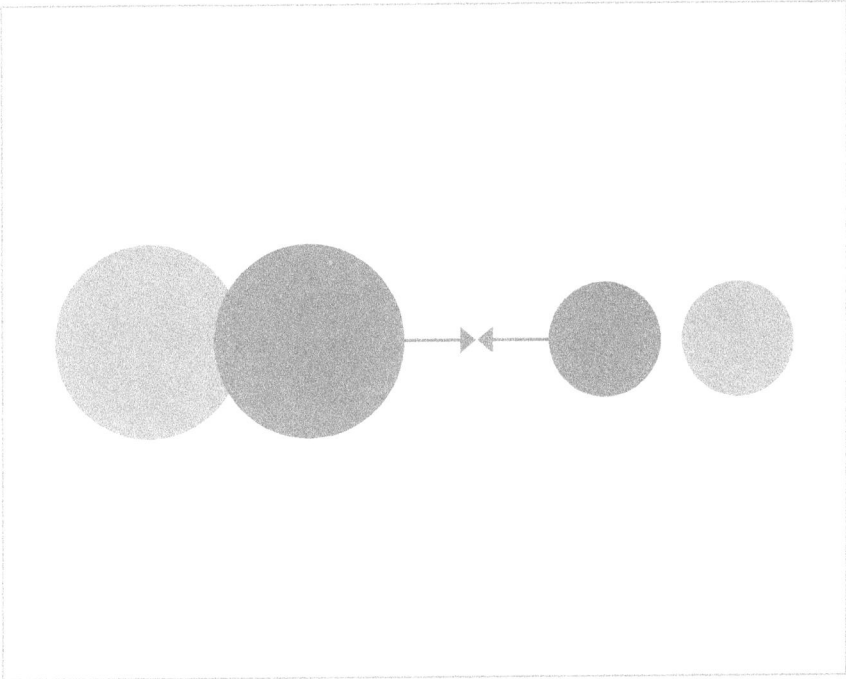

Figure 40.6: This shows the size of the gravitational force acting on the two objects again. They are the same size but in the opposite direction. This is based on what we call the principle of action and reaction in Newtonian dynamics. You may notice that the size of the circular objects on the left and the right are different. If the two sides have the same density but are different sizes, the one on the left is heavier than the one on the right. In other words, they have different mass. Following Newton's second law, the heavier the object is, the smaller the acceleration is given the size of the force. For that reason, the smaller object is going to get accelerated fast. If you are wondering what that is all about, it is time to read the Newton's second law in dynamics part.

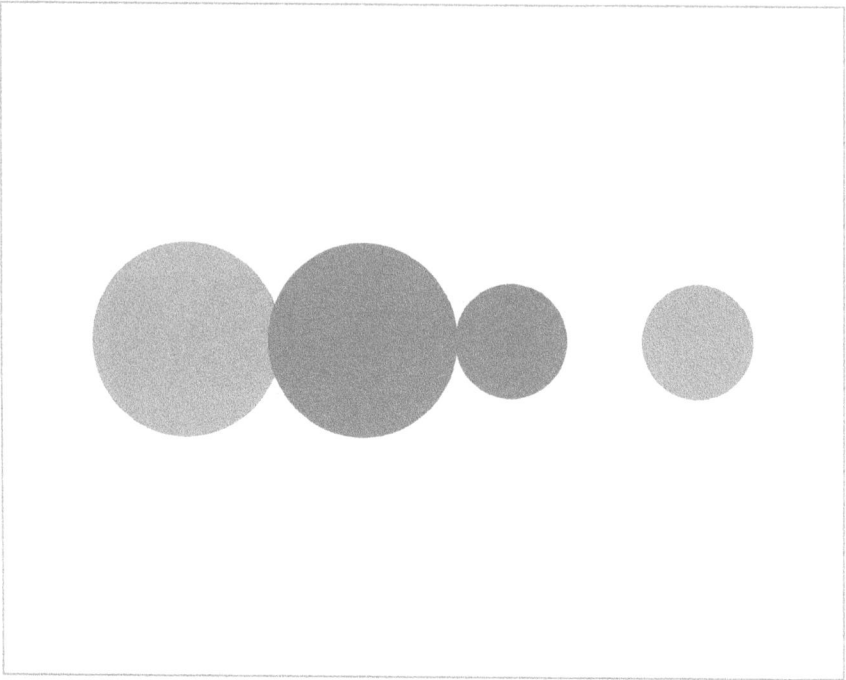

Figure 40.7: Imagine that the two objects colored in light grey are the only two objects in a system and they are going to gravitationally interact with each other. Question: what is going to happen in the end? Answer: Yes, they are going to eventually collide with each other, as shown in grey. In other words, the two objects are going to be in contact momentarily at least. Now, you may wonder how we analyze the case when the two are in contact. That is where we introduce different types of forces such as friction, tension, and normal force. In practice, objects occupy space; they have dimension in space. Thus, we need to think about the cases of their being in contact somewhat differently from those in which they are at a distance.

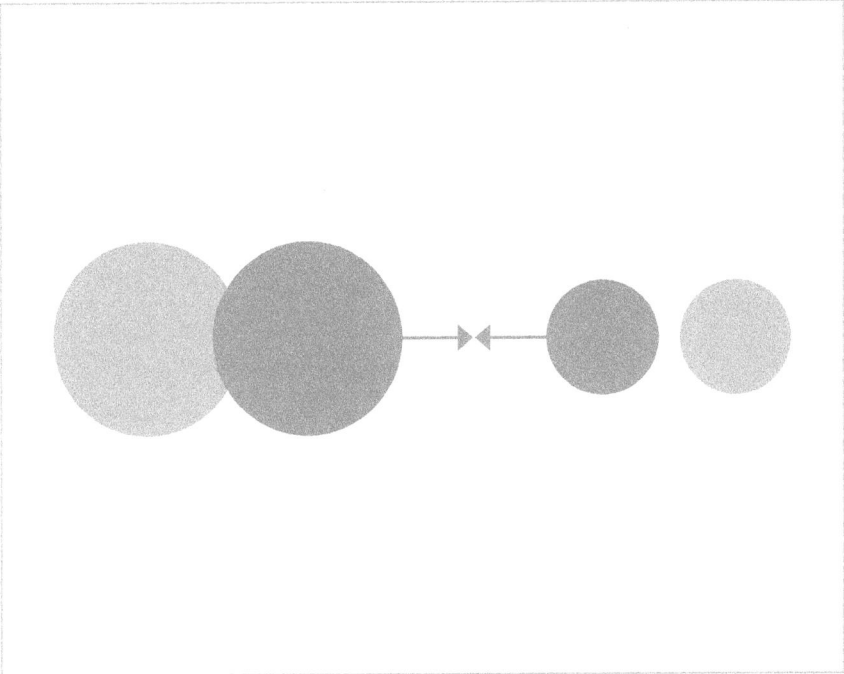

Figure 40.8: The two objects pull each other. Their position as a function of time changes, so we introduce quantities such as velocity and acceleration. After they collide with each other, the mechanical motion associated with them is going to change. How do we quantify that? How do we quantify their kinematics after their collision? We use quantities such as energy and momentum. The total size of momentum is going to be the same before and after their collision. The total size of energy is going to get conserved in an elastic collision, whereas it does not in an inelastic collision. What is that all about? That is all about the conservation of energy and momentum in the dynamics part.

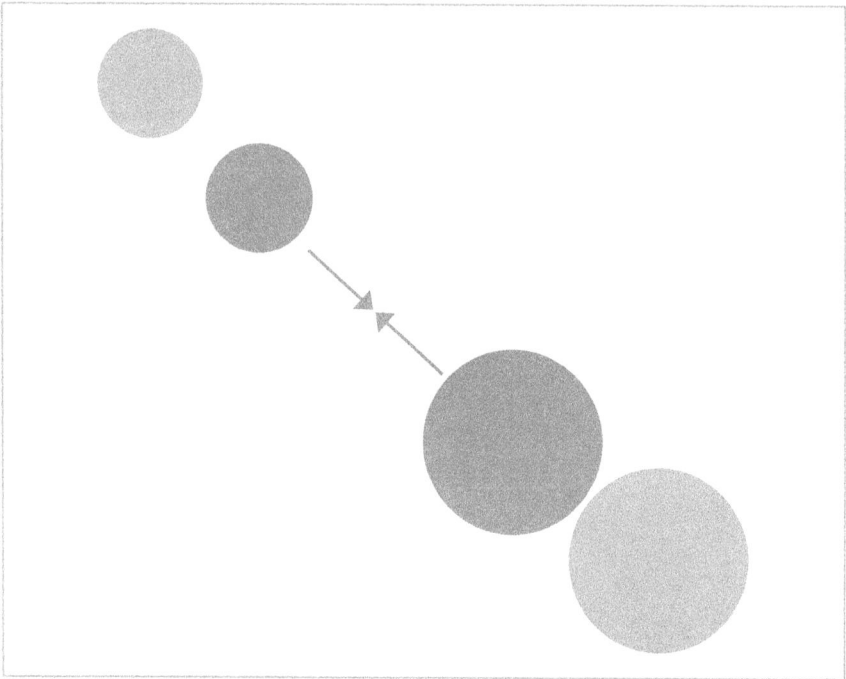

Figure 40.9: This shows everything that we need to know. In summary, when we have more than one object in a system, the objects gravitationally interact with each other, and that causes them to be in motion. There are other causes that could lead them to be in motion too. Their interaction is where everything begins, and the effect associated with the interaction is what we study as kinematics. The cause is what the dynamics part is about. The kinematics part is then to quantitatively describe the mechanical motion; we want to be as specific as possible by introducing quantities such as velocity and acceleration. If it happens in two- or higher dimensional space, that is where the projectile motion gets introduced. Question: think about why we have two spheres colored grey in the figure, why they are colored differently as grey and light grey, what the size of the arrows represents, and why the arrows are the same length.

What do you think? Do you see why we study classical mechanics and what makes it an interesting subject? All the lessons and problems in this book were prepared to describe the gravitational interaction and the motion associated with objects because of the interaction. We just want to understand the cause of the motion and then analyze the motion using some physical quantities. That is all we want to do. Welcome to the introduction to classical mechanics.

Day 41
When you study electrodynamics

"It is not about mass. It is about electric
charge. They are similar but somewhat
different with respect to each other."

Congratulations. You are almost done with reading this book.

In classical mechanics, we study a motion associated with an object or objects with mass. We focus on their having a gravitational interaction and their manifestation in the displacement and time in space and in some variables such as velocity and acceleration that can be derived from them. If you take undergraduate classical mechanics as an elective course but not as part of a physics major, well, in reality, that is all that you may need to remember in practice.

With all probability, you are going to end up taking electromagnetism next if you want to continue studying physics. When you take it, all you need to do first is replace the mass by what is called an electric charge, another type of physical quantity, something that is like what we studied as mass but is somewhat different. Then the rest of what you are going to end up studying is just like what you studied in classical mechanics. For certain, I can tell you that it is going to be a different and interesting subject, but the way that you are going to study it is not that different at all compared to what you did in classical mechanics. For most lessons covered in electromagnetism, you will find a corresponding lesson in classical mechanics.

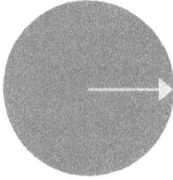

Hi, I am James. I happen to be shaped as a sphere. My radius is 1 inch, and I majored physics.

Sarah: James, I am wondering if you will work at my company. We are looking for someone shaped as a sphere.
Michael: James, I want to hire you because of your 1-inch radius.
James: Am I getting offers for two different reasons?
Sarah and Michael: You as an entity have different characteristics.

Figure 41.1: James can be considered a single entity, but he is getting two different offers from two companies. Again, he is an entity getting two offers because of his having two different properties. How can the conversation illustrated in the figure help us when trying to understand an object interacting with another entity or entities via different types of fundamental interactions?

Problems:

Find a definition of an electric charge. Compare the definition to that of gravitational mass.

Take a look at Figure 41.1. Think about a single entity that has different properties associated with it and so interacts with another entity or entities via different types of fundamental interactions.

Calculate the ratio of the coupling strength of the gravitational interaction and the electromagnetic interaction. Which coupling constant is larger? Write a paragraph on the implication when the strength of the two interactions is comparable with respect to each other.

Epilogue

This was a part of my teaching statement:

"While working as a teaching assistant, I realized that many students taking a general physics course did not work on their homework assignments on their own, but they spent hours searching for the same or similar solutions on the web instead. What did they do with them after that? They just copied and pasted the solutions when submitting their homework. It has been an issue to think hard about for the sake of the physics community.

"Were they copying the solutions just because they were lazy? Or were they just overwhelmed by the amount of work and did not even know where to get started when trying to work on their assignments? Based on my experience, I realized that the case goes to the latter for many students. Furthermore, it has been going on like that for years. In other words, I knew that merely telling them not to do so was not going to work out well. So, in the end, I decided to give it a try and go with a bit of an unconventional approach. In short, I simply gave them the solutions to begin with. The full solution was given to them in the very beginning by writing the entire solution on the board. I showed every single step that was needed to get to the final answer and let students to simply follow mine. In the beginning, they were just very busy with copying and pasting the solution that I provided them, and all they did was follow what I had done.

"But you know what? Things changed slowly and gradually. As the end of the term was getting close, the students, who used to be busy with scribbling down the solutions and not having solid understanding on the course materials and those who used to find solutions on the web, began to think through their homework on their own. Question: why could they complete the assignments if they had not studied as they were supposed to? Answer: They were doing so based on what they copied and pasted in the very beginning. But it was quite amazing to see that the students began to study on their own; they were interested in their studying. They wanted it. They were motivated."

What is the bottom line here? Students just need to build up their basis and need some time to do so. I simply needed to ignite their passion. Then the students were able to think on their own. They needed tools to begin the chain of their thinking. Once they cross the threshold, they work on their own. I learned a lesson there, a huge lesson. I may need to confess that crossing the threshold is not going to be that easy though.

As described in this book, there always is a reason or reasons for an object being in a state of a mechanical motion; if something is in motion, there should be a cause for it. Likewise, if students have a hard time following the course curriculum in a classical mechanics course, there must be a reason or reasons for their having a hard time, and it would be nice to practically deal with them. Again, there always is a reason or reasons for students having a hard time studying physics. It is a matter of identifying the reason or reasons clearly and practically and doing the best to come up with a resolution. This book is a part of it. I truly believe that.

I have been trying to think hard and work hard to develop the ideas and some strategies that could help students so that they have a less challenging time when studying classical mechanics, particularly for those preparing for professional school admissions tests or something equivalent. Writing this book was along the same line of thought. There are many students having a hard time understanding the mathematical parts in physics, and the problem is much worse for students who do not major in physics or mathematics. That is the brutal reality.

Does that mean that they always have a hard time understanding the physics part in physics? I do not think so. In some cases, mathematics can function as a bridge for students to understand physics better and find joy with studying physics later. Then why don't we simply develop another bridge for them to understand the mathematical parts so that they can understand the physics parts better? Or why don't we create material that describes a few important topics in classical mechanics without introducing too many mathematical expressions so that student with some interest in physics can study the mathematical parts on their own later? I really wanted to give it a try at least by writing a book and thus mitigate the issues that many students struggle with one way or the other. I tried to help students from their side.

Is reading this book going to be sufficient for students to gain some insights and a broad range of knowledge in classical mechanics? The answer really depends on what you aim for. If you read this book while taking a classical mechanics course, it is going to help you prepare better. If you read this before taking the course, you will benefit most since you will be prepared to digest the materials being covered in the course. If you read this after taking the course, it will help you understand the materials that you did not understand before. In conclusion, you may need to study further to understand the mathematical side of classical mechanics, but I truly believe that reading this book is going to get you prepared for that and bring passion and curiosity to your mind while studying them further. Thank you for your time.

Jae Jun Kim